Springer Proceedings in Physics 31

Springer Proceedings in Physics

Managing Editor: H. K. V. Lotsch

Volume 30 *Short-Wavelength Lasers and Their Applications*
Editor: C. Yamanaka

Volume 31 *Quantum String Theory*
Editors: N. Kawamoto and T. Kugo

Volume 32 *Universalities in Condensed Matter*
Editors: R. Jullien, L. Peliti, R. Rammal, and N. Boccara

Volume 33 *Computer Simulation Studies in Condensed Matter Physics Recent Developments*
Editors: D. P. Landau, K. K. Mon, and H.-B. Schüttler

Volumes 1 – 29 are listed on the back inside cover

Quantum String Theory

Proceedings of the Second Yukawa Memorial Symposium, Nishinomiya, Japan
October 23–24, 1987

Editors: N. Kawamoto and T. Kugo

With 24 Figures

Springer-Verlag Berlin Heidelberg New York
London Paris Tokyo

Dr. Noboru Kawamoto
Professor Taichiro Kugo, Ph. D.
Department of Physics, Kyoto University, Kyoto 606, Japan

ISBN 3-540-50313-7 Springer-Verlag Berlin Heidelberg New York
ISBN 0-387-50313-7 Springer-Verlag New York Berlin Heidelberg

Library of Congress Cataloging-in-Publication Data. Nishinomiya-Yukawa Memorial Symposium (2nd : 1987 : Nishinomiya-shi, Japan) Quantum string theory. (Springer proceedings in physics; 31) Includes index. 1. String models – Congresses. 2. Quantum field theory – Congresses. I. Kawamoto, N. (Noboru), 1948–. II. Kugo, T. (Taichiro), 1949–. III. Title. QC794.6.S85N57 1987 530.1'4 88-29916

Printing: Weihert-Druck GmbH, D-6100 Darmstadt
Binding: J. Schäffer GmbH & Co. KG., D-6718 Grünstadt
2154/3150-543210 – Printed on acid-free paper

Foreword

My dear friends,

I am very pleased that there are so many participants in the 2nd Nishinomiya-Yukawa Memorial Symposium on Theoretical Physics, and welcome you all sincerely.

Nishinomiya City has been endeavoring to promote education and culture in order to create an attractive town rich in cultural atmosphere. Being proud of the fact that Dr. Hideki Yukawa first created his meson theory in our Nishinomiya City, we are holding the "Nishinomiya-Yukawa Memorial Activities" to promote the study of theoretical physics in honor of Dr. Yukawa's achievements.

The annual activities consist of this symposium, the Nishinomiya-Yukawa Memorial Prize for promising young physicists and Memorial Lectures for citizens. They are run by the Steering Committee, the chairman of which is Professor K. Nishijima, the director of the Research Institute for Fundamental Physics, Kyoto University. I would like to express my sincere gratitude to the members of this committee for their great efforts, as well as to the members of the Organizing Committee of this symposium.

I am very happy that this first-class international symposium is being held at Nishinomiya, as is fitting for this town of education that encourages the spirit of pursuing the truth. It will be a great honor if this symposium is successful and fruitful for many participants, particularly for young physicists.

Thank you

Yoneji Yagi
Mayor of Nishinomiya

Preface

This volume contains lectures presented at the second Nishinomiya-Yukawa Memorial Symposium on "Quantum String Theory" held on October 23 and 24, 1987 in Nishinomiya City. To our regret, however, two important contributions by A.M. Polyakov and D. Friedan are missing.

The realistic possibility of achieving the long-standing dream of unifying all elementary interactions has struck many physicists and encouraged them to work on strings in recent years. The purpose of the present symposium was, thus, to bring together the main contributors to this renewed exciting subject, to discuss the latest developments, and hopefully to gain perspectives for the future.

We were pleased to have outstanding foreign guests as speakers, who attracted many Japanese physicists active in this field. The atmosphere of the symposium, like that of a subsequent workshop on "Superstrings and Related Topics" held October 26–30, 1987 at the Research Institute for Fundamental Physics, Kyoto University, was excellent, with many intense and enlightening discussions.

The organizing committee of the symposium consisted of
Y. Fujii (University of Tokyo)
A. Hosoya (Hiroshima University)
N. Kawamoto (Kyoto University)
K. Kikkawa (Osaka University)
M. Kobayashi (KEK)
T. Kugo (Kyoto University)
K. Nishijima (Kyoto University)
A. Tomimatsu (Nagoya University)

The symposium was organized under the auspices of the Education Board of Nishinomiya City, and the Research Institute for Fundamental Physics, Kyoto University.

Kyoto, July 1988

Noboru Kawamoto
Taichiro Kugo

It was a shock to all of us to learn that Vadik Knizhnik passed away on December 15, 1987 at the age of 25, just two months after the symposium. We were all impressed by the brilliant young physicist who made a great contribution to the symposium and had warm contacts with many people (see picture). We dedicate this volume to his memory.

Vadik Knizhnik (*back center*), together with A.A. Belavin (*back left*)and young Japanese students.

Contents

Part I

String Theories: General Aspects

Uniqueness of Superstrings

L. Brink

Institute of Theoretical Physics, S-412 96 Göteborg, Sweden

1. INTRODUCTION

During the last 15 years elementary particle physics has undergone a real revolution. Concepts that had been debated for a long time are now parts of the basic building blocks. The most obvious of these is relativistic quantum field theory. All through history from Faraday onwards its relevance has been questioned, up to the last two decades. Now gauge field theories have proved to be extremely successful and no one really doubts that this is the correct formalism to use to describe the strong, the weak and the electromagnetic interactions, at least at the energies presently accessible. The problem now with these theories is really on the philosophical side. We know that there are (seemingly) a huge class of gauge field theories that could describe a universe, each one endowed with a large number of free parameters. Is this what we can expect from physics? I would hope not!

A more appealing idea to me is that there exists only one fully consistent theory and that the gauge field theories we believe in are some kind of low energy effective theories. This idea is, of course, behind the unification idea. We have seen a successful unification of the weak and the electromagnetic interactions and there are also clear signs of their unification with the strong interaction. The failure in this approach is that it does not give any hints as to how to include also gravity, which should be done since at the unification energies it cannot be neglected. An alternative route is then to start with gravity and to try to extend it so as to include also the other interactions. This led to the supergravity models [1] in which gravity could merge with gauge field theories in a geometric way. However, supergravity models show the same reluctance to be quantized as ordinary gravity. In fact, it is most reasonable to conclude that it is impossible to quantize these models, at least using standard particle physics procedures [2].

At this stage it is appropriate to return to the starting point (which we claimed led to all the glorious results), quantum field theory. The field theories we have considered are field theories for point-like objects since we use local field theories. The lesson we have learnt is that there are whole

Springer Proceedings in Physics Vol. 31: **Quantum String Theory**
Editors: N. Kawamoto and T. Kugo

classes of renormalizable gauge field theories which in principle could describe the strong, the weak and the electromagnetic interactions, while there are (seemingly) no finite or renormalizable theories including gravity.

One possible way out of this dilemma is to try to change the starting point but in a way that will not ruin the successes at our present day energies with point-particle theories. It is then natural to question whether we could use extended objects with a size that is negligible at present day length scales. However, such objects, regardless of size, usually lead to causality breaking theories and were therefore in the past more or less abandoned. However, from attempts to construct hadronic models, string theories emerged [3], theories with one-dimensional objects. These models avoided all the usual problems and furthermore they have some properties which make them very attractive. First of all it was found that the number of models is quite limited and their structure very restricted. Secondly they necessarily contain a spin-2 particle coupling appropriately to be interpreted as a graviton.

In this lecture I will discuss the various possible ways to construct string models and how they might lead to a (unique) unified model of all interactions.

2. BOSONIC STRINGS

We start by considering a string which is completely described by its coordinate $x^{\mu}(\sigma,\tau)$ where the world-sheet spanned out by the propagating string is parametrized by a time-like variable τ and a space-like one σ. The latter is chosen to run between 0 and π. Although we will start by considering a free string, we must construct a free theory which could interact in a natural way. (In fact, one lesson we have learnt from string theory is that in consistent models, interactions are essentially built in if we allow the free strings to split and join).

Since we will describe extensions of gravity theory it would have been natural to start somehow with gravity and take off from there. However, no such path has so far been successful and we will instead start from freely propagating strings and consider their possible interactions. We will return in the end to this question. The interacting theories that we aim at should satisfy

i) Poincaré invariance
ii) causality
iii) unitarity.

In the standard treatment the Poincaré invariance is automatic; the unitarity is ensured by the gauge invariance of the theory, the reparametrization invariance, and causality is implemented by choosing the correct description of the interactions. There is, however, an alternative route, which will quickly show us the various possibilities, but will obscure the underlying geometry. This one we will follow here. The basic problem for interacting extended objects alluded to above is that different parts carry different times in general. If this is so there seems to be no way to set up a consistent causal interaction. To avoid this problem we simply decree that each point along the string carry the same time. We do have a choice here, since, as Dirac has shown [4], any time-like variable can be used as the evolution parameter; in fact we can even use the limiting case of one of the light-cone directions $x^{\pm} = \frac{1}{\sqrt{2}}(x^0 + x^{d-1})$. (The Minkowski space has d dimensions). Since we are going to describe a massless string, i.e. a string with no explicit mass term in the free action, it is natural to choose x^+ as the "time". Our first assumption is then (we use the standard notation $\dot{x} \equiv \frac{\partial x}{\partial \tau}$ and $x' \equiv \frac{\partial x}{\partial \sigma}$)

$$x'^{+} = 0 \quad , \tag{2.1}$$

i.e.

$$x^{+}(\sigma,\tau) = x^{+} + p^{+}\tau \quad , \tag{2.2}$$

where x^+ and p^+ are integration constants. We will interpret p^+ as the total momentum in the +-direction. In order to do so we have put a constant with dimensions to one. We will soon correct this fact.

To implement the third requirement above, we assume that all the propagating degrees of freedom lie in the transverse part of $x^{\mu}(\sigma,\tau)$, i.e. in $x^i(\sigma,\tau)$, where $i = 1,\ldots,d-2$. Since we are going to describe freely propagating strings, the natural action is

$$S = -\frac{T}{2}\int_{\tau_i}^{\tau_i} d\tau \int_0^{\pi} d\sigma\, \eta^{\alpha\beta}\, \partial_\alpha x^i \partial_\beta x^i \quad , \tag{2.3}$$

with α and β taking the values τ and σ and the metric $\eta^{\alpha\beta}$ is chosen space-like. T is a propertionality factor (which ensures that x^i is a length) and can be understood to be the string tension.

The possible flaw with this action is that it is not explicitly Poincaré invariant. The invariance can only be proven by an explicit construction of the Poincaré generators. Before performing the analysis let us check the dynamical content of the action. When varying it to obtain the equations of motion some care has to be exercised because of the finite σ-inverval. Disregarding possible surfaces terms, one obtains the equation of motion

$$\ddot{x}^i - x''^i = 0 \quad . \tag{2.4}$$

The canonical conjugate momentum density is

$$p^i(\sigma,\tau) = T\dot{x}^i(\sigma,\tau) \quad , \tag{2.5}$$

and the total momentum is

$$p^i = T \int_0^\pi d\sigma \, \dot{x}^i(\sigma,\tau) \quad . \tag{2.6}$$

By comparing with (2.2) we find that by choosing $T = \frac{1}{\pi}$, also p^+ defined by (2.2) satisfy (2.6). For the rest of the lectures we will make this choice.

We can now quantize the theory by the canonical commutators (with $\hbar = 1$).

$$[x^i(\sigma,\tau), \, p^j(\sigma',\tau)] = i\, \delta^{ij}\delta(\sigma - \sigma') \quad . \tag{2.7}$$

Returning to the surface terms we find two sets of boundary conditions which make the surface terms vanish:

i) $x^i(\sigma = 0,\pi) = 0$, open strings:

$$x^i(\sigma,\tau) = x^i + p^i\tau + i \sum_{n\neq 0} \frac{1}{n} \alpha_n^i \cos n\sigma \, e^{-in\tau} \quad , \tag{2.8}$$

ii) x^i periodic , closed strings:

$$x^i(\sigma,\tau) = x^i + p^i\tau + \frac{i}{2} \sum_{n\neq 0} \frac{1}{n} (\alpha_n^i e^{-2in(\tau-\sigma)} + \tilde{\alpha}_n^i e^{-2in(\tau+\sigma)}) \quad . \tag{2.9}$$

The canonical commutators are satisfied if

$$[\alpha_n^i, \alpha_m^j] = n\, \delta_{n+m,0}\, \delta^{ij} \tag{2.10}$$

$$[\tilde{\alpha}_n^i, \tilde{\alpha}_m^j] = n\, \delta_{n+m,0}\, \delta^{ij} \quad . \tag{2.11}$$

The Hilbert space of states can be constructed by introducing a vacuum $|0\rangle$ and demanding

$$\alpha_n^i \, |0\rangle = 0 \qquad n > 0$$

$$\tilde{\alpha}_n^i \, |0\rangle = 0 \qquad n > 0 \quad . \tag{2.12}$$

The space of states is then constructed by letting α_{-n}^i and $\tilde{\alpha}_{-n}^i$, $n > 0$, act as creation operators building up an explicitly positive-norm Hilbert space.

It remains to check the Poincaré invariance. The generators must be constructed out of $x^i(\sigma)$, $p^i(\sigma)$, x^+, p^+ and possibly a zero mode x^- conjugate to p^+. Since there is no such linear Lorentz transformation, they must be <u>non-linear</u>. We will first construct the representation of the generators classically and use Poisson brackets instead of the commutators (2.7). We will also use a notation with lowercase letters for these generators and use capital letters for second-quantized representations. In the light-cone frame the generator p^- is the Hamiltonian. Now

$$p^- \sim \frac{\partial}{\partial x^+} = i\,\frac{\partial\tau}{\partial x^+}\frac{\partial}{\partial\tau} = \frac{1}{p^+}\left(i\,\frac{\partial}{\partial\tau}\right) \sim \frac{1}{p^+}\,H \quad , \tag{2.13}$$

where H is the Hamiltonian connected to the Lagrangian (2.3). The Hamiltonian density is

$$H = \dot{x}^i p^i - \mathcal{L} = \frac{1}{2\pi}\left[\pi^2 p^{i^2} + x'^2\right] \quad . \tag{2.14}$$

Hence the translation part of the algebra is

$$p^- = \frac{1}{2\pi p^+}\int_0^\pi d\sigma\left[\pi^2 p^{i^2} + x'^{i^2}\right] \tag{2.15}$$

$$p^+ = p^+ \tag{2.16}$$

$$p^i = \int_0^\pi d\sigma\, p^i(\sigma) \quad . \tag{2.17}$$

The Lorentz part we write as

$$j^{ij} = \int_0^\pi d\sigma\,(x^i p^j - x^j p^i) \tag{2.18}$$

$$j^{+i} = \int_0^\pi d\sigma\,(x^+ p^i - x^i p^+) \tag{2.19}$$

$$j^{+-} = x^+ p^- - x^- p^+ \tag{2.20}$$

$$j^{-i} = \int_0^\pi d\sigma\left[x^-(\sigma)p^i(\sigma) - x^i(\sigma)p^-(\sigma)\right] \quad . \tag{2.21}$$

Here we have introduced $x^-(\sigma)$ which should be determined in terms of x^i and p^i apart from its zero mode x^-. We have introduced the generators one by one. The generator j^{ij} follows from the explicit transverse invariance in (2.3). The generators j^{+i} and j^{+-} are natural and are easily seen to be correct. Finally we determine j^{-i} first classically (to avoid ordering problems). A tedious calculation shows that the algebra is indeed satisfied if we choose

$$\acute{x}^{-}(\sigma) = \frac{\pi}{p^{+}} p^{i} x^{i} \quad . \tag{2.22}$$

If we so check the quantum algebra we symmetrize the generators in order to keep the hermiticity. This affects the checking of the commutator $[j^{i-}, j^{j-}] = 0$, since j^{i-} is cubic. A detailed computation shows that the commutator is satisfied only if d = 26 (the critical dimension) [5].

Inserting the solution (2.8) into p^{-} (2.15) we obtain

$$p^{-} = \frac{p^{i^2}}{2p^{+}} + \frac{1}{p^{+}} \sum_{n=1}^{\infty} \alpha_{-n}^{i} \alpha_{n}^{i} \quad , \tag{2.23}$$

i.e.

$$p^{2} = -2 \sum_{n=1}^{\infty} \alpha_{-n}^{i} \alpha_{n}^{i} = -m^{2} \quad . \tag{2.24}$$

Reintroduce the non-dimensionless constant as $\alpha' = \frac{1}{2\pi T}$, the Regge slope. Then the correct expression is

$$\alpha' m^{2} = \sum_{n=1}^{\infty} \alpha_{-n}^{i} \alpha_{n}^{i} \quad . \tag{2.25}$$

The squared mass is hence built up by an infinite set of "harmonic oscillator energies" [6]. For the quantum case each oscillator will contribute an energy due to zero-point fluctuations (which is due to the symmetrization above). The lowest mass level will then be

$$\alpha' m_{0}^{2} = \frac{d-2}{2} \sum_{n=1}^{\infty} n \tag{2.26}$$

This is clearly a divergent sum which must be regularized. We do so by comparing the sum to the Riemann ζ-function [7]

$$\zeta(s) = \sum_{n=1}^{\infty} n^{-s} \quad , \quad \mathrm{Re}\, s > 1 \quad . \tag{2.27}$$

This is a function that can be analytically continued to s = -1 and

$$\zeta(-1) = -\frac{1}{12} \quad . \tag{2.28}$$

In this way the infinite series has been regularized into

$$\alpha' m_0^2 = -\frac{d-2}{24} \quad . \tag{2.29}$$

Hence this term should be included in p^-.

There is also a more standard way [6] of obtaining this result by adding counterterms to the action (2.3), which amounts to renormalizing the speed of light, (which, of course, is another parameter of the theory). It is quite attractive that it is the quantum theory that demands a finite speed of light.

The really important consequence of Eq. (2.29) is that the lowest state is a tachyon. This really means that an interacting theory based on this bosonic string will not make sense. Also since the tachyon is the scalar state with no excitations of the higher modes, I find it hard to believe that there exists a consistent truncation in which the tachyon is left out.

Let us now redo the last analysis for closed strings. Inserting the solution (2.9) we find

$$\frac{\alpha'}{4} m^2 = \sum_{n=1}^{\infty} (\alpha_{-n}^i \alpha_n^i + \tilde{\alpha}_{-n}^i \tilde{\alpha}_n^i) \equiv N + \tilde{N} \quad . \tag{2.30}$$

Also in this sector we find a tachyon. For closed strings we get a further constraint. Consider again Eq. (2.22). Integrating it between 0 and σ we find $x^-(\sigma)$. Although $x^-(\sigma)$ depends on the x^i's we must demand that it represent a component of $x^\mu(\sigma)$ and hence must be periodic. Then

$$\int_0^\pi d\sigma\, x'^- = 0 = \frac{1}{p^+} \int_0^{0} d\sigma\, \dot{x}^i x'^i = \frac{\pi}{p^+} (N - \tilde{N}) \quad , \tag{2.31}$$

i.e. classically $N = \tilde{N}$ and quantum mechanically we impose this condition on the physical states. Note that this implies

$$\frac{\alpha'}{4} m^2 = 2N = 2\tilde{N} \quad . \tag{2.32}$$

The formulation derived here is the "light-cone gauge" formulation of the Nambu-Goto string action [8]. In the usual description [5] one starts with a reparametrization and Poincaré invariant action. By specifying a gauge, one arrives at the formulation above. An approach which is certainly more general starts with a geometric action. This is quite different from the approach here, where we start with a non-covariant action which, however, explicitly describes free positive-norm states.

The Poincaré algebra spanned by the generators (2.15)-(2.21) with the constraint (2.22) contains all the information about the strings. This is typical for the light-cone gauge. By finding the non-linear representation we know the complete dynamics of the system, since the Hamiltonian is one of the generators. The regrettable thing is that so far we have not found a deductive way to find the generators, but have had to allow some guesswork and then check.

By describing the dynamics of free bosonic strings we have found that such a theory is much more constrained than a corresponding theory for point particles, for which there are no obstacles like above. This is certainly a most wanted property and might indicate that we are on our way towards a more unique description of Nature.

3. SUPERSTRINGS

The representation of the Poincaré algebra found in the last section was found to lead to an inconsistent theory. To obtain a consistent one, we need to change the expression for p^- (2.15) so as to avoid tachyons. The natural thing is to introduce a set of anticommuting harmonic oscillators such that their zero-point fluctuations compensate the ones from the commuting oscillators.

To get a full compensation we need anticommuting oscillators transforming as d-2 under SO(d-2). For lower d's there might be more than one such representation. We can always try the vector representation, but for d=3, 4, 6 and 10 we can also use spinor representations. Here we will use a spinor representation but later we will see that the vector representation leads to the same model.

Consider hence a set of anticommuting oscillators S_n^a, a = 1,...,8 satisfying

$$\{S_m^a, S_n^b\} = \delta_{m+n,0}\,\delta^{ab} \quad . \tag{3.1}$$

If the relevant mass formula for open strings is

$$\alpha' p^2 = -\left\{\sum_{n=1}^{\infty} \alpha_{-n}^i \alpha_n^i + \sum_{n=1}^{\infty} n S_{-n}^a S_n^a\right\} \quad , \tag{3.2}$$

we can deduce p^- from this expression. To write it in a coordinate basis we introduce two normal-mode expansions

$$S^{1a} = \sum_{n=-\infty}^{\infty} S_n^a\, e^{-in(\tau-\sigma)} \tag{3.3}$$

$$S^{2a} = \sum_{n=-\infty}^{\infty} S_n^a\, e^{-in(\tau+\sigma)} \quad , \tag{3.4}$$

such that

$$\{S^a_A(\sigma,\tau),\ S^b_B(\sigma,\tau)\} = \pi\ \delta_{AB}\ \delta^{ij}\ \delta(\sigma-\sigma') \quad . \tag{3.5}$$

The generator p^- can then be written as

$$p^- = \frac{1}{2\pi p^+} \int_a^\pi (\pi^2 p^{i^2} + x'^{i^2} - iS^{1a}S'^{1a} + iS^{2a}S'^{2a}) \quad . \tag{3.6}$$

Before trying to construct the remaining generators let us consider the dynamics following from (3.6). Since $p^- = \frac{1}{p^+} H$, the corresponding action is

$$S = -\frac{1}{2\pi} \int d\tau d\sigma\ [\eta^{\alpha\beta} \partial_\alpha x^i \partial_\beta x^i + i\ \bar{S}^a \sigma^\alpha \partial_\alpha S^a] \quad , \tag{3.7}$$

where we combine the two S^a's into two-component spinors and use the Majorana representation for the 2×2 Dirac matrices σ^α. When going from the Hamiltonian to the action we have assumed that surface terms vanish when S is varied. In the case of open strings the possible boundary conditions for the fermionic fields are

$$S^{1a}(0,\tau) = S^{2a}(0,\tau) \tag{3.8}$$

$$S^{1a}(\pi,\tau) = \begin{cases} S^{2a}(\pi,\tau) & (3.9a) \\ -S^{2a}(\pi,\tau) & (3.9b) \end{cases} \quad .$$

In the solution (3.3)-(3.4) we have in fact used (3.8) and (3.9a). The other choice leads to half-integer moded solutions and this will again give a tachyonic state, so it must be discarded.

We can now build states with the bosonic modes α^i_n satisfying (2.12) and S^a_n satisfying

$$S^a_n\ |0\rangle = 0 \qquad n > 0 \quad . \tag{3.10}$$

It is easy to see that the spectrum contains both bosons and fermions and the elimination of the possible other sector of the model is not harmful. If we had chosen instead to use a vector representation for the anticommuting modes we must use both sectors and impose truncations such that the tachyon disappears. This is the Ramond-Neveu-Schwarz formulation [9] and the truncation is the Gliozzi-Olive-Scherk projection [10].

The action (3.7) also allows for periodic boundary conditions and anti-periodic boundary conditions. The latter ones will lead to tachyons and are hence discarded. The solution for closed strings is then (2.9) and

$$S_1^a = \sum_{n=-\infty}^{\infty} S_n^{1a} e^{-2in(\tau-\sigma)} \tag{3.11}$$

$$S_2^a = \sum_{n=-\infty}^{\infty} S_n^{2a} e^{-2in(\tau+\sigma)} \quad . \tag{3.12}$$

It remains now to show that the models following from (3.7) indeed are Poincaré invariant. This is an arduous task which can only be done by some guesswork. The marvellous result is that the action in fact carries an even bigger symmmetry! [11] It is super-Poincaré invariant. Since light-cone supersymmetry might not be too familiar, let me first briefly review it in d=10 (which is the dimension we will use). The supersymmetry charge Q decomposes into two SO(8) light-cone spinors Q_+^a and $Q_-^{\dot a}$, where the indices $a,\dot a = 1,\ldots,8$ denote the two inequivalent 8-component spinors of SO(8). The algebra is

$$\{Q_+^a, Q_+^b\} = 2p^+\delta^{ab} \tag{3.13a}$$

$$\{Q_-^{\dot a}, Q_-^{\dot b}\} = 2p^-\delta^{\dot a\dot b} \tag{3.13b}$$

$$\{Q_+^a, Q_-^{\dot b}\} = \sqrt{2}(\gamma^i)^{a\dot b}p^i \quad . \tag{3.13c}$$

The most general super-Poincaré algebra is now written for closed strings and carry an N=2 supersymmetry. The generators are

$$p^+ = p^+ \tag{3.14a}$$

$$p^i = \int_0^\pi d\sigma\, p^i(\sigma,\tau) \tag{3.14b}$$

$$p^- = \frac{1}{2\pi p^+}\int_0^\pi d\sigma\, [\pi^2 p^{i\,2} + x'^{i\,2} - i(S^1 S'^1 - S^2 S'^2)] \tag{3.14c}$$

$$q_1^{+a} = \sqrt{\frac{2p^+}{\pi}}\int_0^\pi d\sigma\, S_1^{\,a} \tag{3.15a}$$

$$q_2^{+a} = \sqrt{\frac{2p^+}{\pi}}\int_0^\pi d\sigma\, S_2^{\,a} \tag{3.15b}$$

$$q_1^{-\dot{a}} = \frac{1}{\pi\sqrt{p^+}} \int_0^\pi d\sigma (\gamma^i S_1)^{\dot{a}} (\pi p^i - x'^i) \tag{3.15c}$$

$$q_2^{-\dot{a}} = \frac{1}{\pi\sqrt{p^+}} \int_0^\pi d\sigma (\gamma^i S_2)^{\dot{a}} (\pi p^i + x'^i) \tag{3.15d}$$

$$j^{ij} = \int_0^\pi d\sigma [x^i p^j - x^j p^i + \frac{1}{4\pi} (S^1 \gamma^{ij} S^1 + S^2 \gamma^{ij} S^2)] \tag{3.16a}$$

$$j^{+i} = \int_0^\pi d\sigma (x^i p^j - x^i p^+) \tag{3.16b}$$

$$j^{+-} = x^+ p^- - x^- p^+ \tag{3.16c}$$

$$j^{-i} = \frac{1}{2} \int_0^\pi d\sigma [\{x^-(\sigma), p^i\} - \{x^i, p^-(\sigma)\}$$

$$- \frac{1}{4\pi\sqrt{\pi p^+}} (S^1 \gamma^{ij} S^1 (\pi p^j - x'^j) + S^2 \gamma^{ij} S^2 (\pi p^j + x'^j)) + 4 \frac{p^i}{p^+}] \quad , \tag{3.16d}$$

where

$$x'(\sigma) = \frac{\pi}{p^+} p^i x'^i + \frac{i}{2p^+} (S^1 S'^1 + S^2 S'^2) \quad . \tag{3.17}$$

If the dynamical variables are quantized and care is exercised so that the generators are hermitean then again we find that it only works in a critical dimension, namely in d=10.

This is, in fact, the most general algebra we can construct which can be used for realistic string theories. We could attempt to increase the number of bosonic vectors, but that leads to models with critical dimensions ≤ 2 [12].

The critical generator, which determines the critical dimension is always j^{-i}. In order for it to close properly we need

8 bosonic coordinates $x^i(\sigma,\tau)$

8 fermionic coordinates $S_A^a(\sigma,\tau)$.

However, we only use their local properties. Nowhere have we used that the bosonic coordinates go from $-\infty$ to $+\infty$. In fact a bosonic coordinate can very well be a small cyclic one. That would mean that the SO(9,1) symmetry is broken, but who cares? In the end we only want an SO(3,1) symmetry. If six of the coordinates are cyclic we would have such a symmetry as long as j^{-i} (i=1,2) were the ones of (4.12d).

A further freedom in the algebra comes from the fact that every generator is split into one term with right-moving coordinates $x^i(\tau-\sigma)$ and $S_1{}^a(\tau-\sigma)$ and one with left-moving coordinates $\tilde{x}^i(\tau+\sigma)$ and $S_2{}^a(\tau+\sigma)$, which completely commute with each other. This allows us to make various combinations. The best known example of this is the heterotic string [13] which combines a right-moving superstring with a left-moving bosonic one. This model is still tachyon-free since $m^2 = 8N$ (as in (2.32)).

We see here vast possibilities in the construction of string models and we need further constraints to cut down the number of possible models. One such constraint is modular invariance, which we turn to now.

4. MODULAR INVARIANCE

Before checking that consistent interactions can be set up, there is one test we can do on our models. Every particle that can propagate will give a contribution to the cosmological constant. To lowest order this contribution is essentially the trace of the propagator. For a string the world-sheet of this trace is (after a Wick-rotation $\tau \to it$) a torus. A torus can be written as R^2/Γ^2 where Γ^2 is a two-dimensional lattice (see Fig. 1)

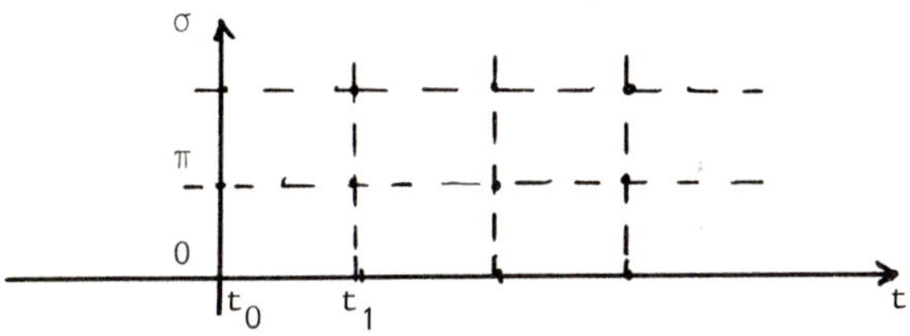

Fig. 1

A torus can be characterized by a complex number (most often) called τ. To obtain the cosmological constant we must integrate over all tori, i.e. over τ. The integrand is then a function $f(\tau)$. It is obtained by considering a propagation in the t-direction with periodic boundary conditions. Consider now the same computation in the σ-direction [14]. We know from before that the Hamiltonian is symmetric in σ and τ and since also in this direction we have periodic boundary conditions (we consider closed strings) we must obtain the same result. The interchange of σ and t amounts to let $\tau \to -\frac{1}{\tau}$ for the torus and hence the integrand should be invariant under this change.

Furthermore we can consider the propagation from $0 \leq \sigma \leq \pi$ at t_0 to $\pi \leq \sigma \leq 2\pi$ at t_1. This computation must be identical again to the ones above since the initial and final states are the same. Again the integrand must show this invariance which corresponds to a shift $\tau \to \tau + 1$.

Combining the two transformation we can write them as

$$\tau \to \frac{a\tau + b}{c\tau + d} \quad , \quad ad - bc = 1 \quad , \tag{4.1}$$

where a,b,c and d are integers. These transformations are called modular transformations and the group is called the modular group.

The integrand in the cosmological constant is essentially a partition function enumerating the multiplicity of states at each mass level. The requirement of modular invariance amounts to constraints on the spectrum of physical states, and it will be a strong constraint on possible models.

For supersymmetric theories the cosmological constant is zero and for the cases where only one diagram contributes, the above exercise is trivial. Then one must consider ordinary loop diagrams. However, for most models different contributions are added and then the discussion above is applicable.

5. POSSIBLE STRING MODELS

Let so return to the super-Poincaré algebra and discuss the various possibilities we obtain by different choices for the boundary conditions for the coordinate.

i) _Type IIb Superstring._ This is the full algebra (4.10)-(4.12). It is hence a 10-dimensional model and the massless sector is type IIb supergravity, which is a chiral model since the creation operators S^{1a}_{-n} and S^{2a}_{-n} create spinors of only one chirality. There is no Yang-Mills multiplet and in fact no vector particle at all, which makes this model inadequate for a direct compactification of six of the coordinates.

ii) _Type IIa Superstring._ The anticommuting coordinate S^{a}_{2} can instead be chosen to transform as the other spinor representation, $S^{\dot{a}}_{2}$. Nothing is affected in the algebra since S_1 and S_2 are never contracted with each other. This is then a 10-dimensional non-chiral model with type IIa supergravity as the massless sector. The non-chirality makes this model inadequate.

Both these models satisfy the modular invariance. If we had used the formulation with an anticommuting coordinate transforming as a vector under SO(1,9), the modular invariance demands the truncation mentioned above.

iii) _Type I Superstring._ For open superstrings we must use the boundary conditions corresponding to (2.8) and (3.8) and (3.9a). Then $q_1^{+} = q_2^{+}$ and $q_1^{-} = q_2^{-}$ and the supersymmetry is reduced to an N=1 one. One can also perform this truncation for closed strings. For the open strings whose massless sector is a Yang-Mills multiplet we can allow for an internal (global) symmetry (the remnant of the gauge symmetry in a covariant formalism). What one does is to associate a Gell-Mann matrix (λ_i) with the i'th external string state and multiply an N-point scattering amplitude with a group theory factor $tr(\lambda_1 \dots \lambda_N)$.

This amounts to letting each state transform as $\Phi^a{}_b$, where the indices a and b run over the fundamental representation and its complex conjugate. The intuitive way to interpret this scheme is to say that each end of the string carries a "quark" and "antiquark", respectively.

The cosmological constant for this model is zero because of the supersymmetry, but a one-loop computation of the interacting theory shows that a finite result is only obtained when the gauge group is SO(32) [15]. The uniqueness of this group was originally found by Green and Schwarz when they showed that only for this gauge group is the model anomaly free at the one-loop level [16].

The model is 10-dimensional, has chiral fermions and it is meaningful to investigate a compactification to 4 dimensions. No such scheme has, however, looked promising, mainly since the symmetry SO(32) seems hard to break to a realistic gauge group.

iv) Heterotic string. We pointed out earlier that we can combine left-going modes from one model with right-going ones from another. In the heterotic string one combines a right-moving superstring with a left-moving bosonic string. We can then only demand an SO(1,9) symmetry at most. What happens to the 16 extra left-moving coordinates? Such a coordinate can be written as

$$x^I(\sigma,\tau) = x^I + p^I(\tau+\sigma) + \frac{i}{2} \sum_{n\neq 0} \frac{\tilde{\alpha}_n^I}{n} e^{-2in(\tau+\sigma)} \quad . \tag{5.1}$$

However, it is not consistent with periodic boundary conditions in σ! To solve this dilemma we are forced to allow for more general boundary conditions. If the coordinates x^I lie on a hypertorus with radius R^I then the function $x^I(\sigma,\tau)$ maps the circle $0 \leq \sigma \leq \pi$ onto the circle $0 \leq x^I \leq 2\pi R^I$, and such maps fall into homotopy classes characterized by a winding number L^I that counts how many times x^I wraps around the circle. Then (5.1) is an allowed solution with $p^I = 2L^I R$. We are hence forced to consider the extra 16 coordinates x^I to span a hypertorus. This is physically appealing since such a hypertorus can be small and we do not need to worry about X^I on a macroscopic scale.

For a coordinate on a hypertorus the momentum is quantized and spans a space $R^{16}/_{\Gamma^{16}}$, where Γ^{16} is the lattice describing the hypertorus. Γ^{16} and $\tilde{\Gamma}^{16}$ are dual to each other, which amounts to saying the basis vectors in the two lattice lattice are orthonormal to each other. From the conditon on the momentum above we see that $\tilde{\Gamma}^{16} \quad \Gamma^{16}$.

As was pointed out in (2.32) the mass of a state depends only on the mode number N of one of the sets of modes. Since the heterotic string has a right-moving sector of a superstring, there are no zero-point fluctuations in this sector and hence the lowest state of the heterotic string is massless.

If we so impose the constraint of modular invariance we will find that the lattices Γ^{16} and $\tilde{\Gamma}^{16}$ must satisfy the constraint $\Gamma^{16} = \tilde{\Gamma}^{16}$, i.e. the lattice must be self-dual. This is a very strong restriction and in 16 dimensions there are only two such lattices, the root lattice of $E_8 \times E_8$ and a weight lattice of Spin $(32)/_{Z_2}$ (which at the algebra level is SO(32).)

The most remarkable thing now is that the 16 coordinates x^I carries a rank 16 symmetry group. In a point-particle case we would expect an SO(16) symmetry, but here the possible symmetry groups are the above-mentioned groups $E_8 \times E_8$ or Spin $(32)/_{Z_2}$.

The heterotic string is then a d=10 model where the massless sector is N=1 supergravity coupled to an N=1 super-Yang-Mills multiplet with an $E_8 \times E_8$ gauge group (or Spin $(32)/_{Z_2}$). These two E_8's only communicate via gravitational interactions so the gauge group to investigate is really E_8. This is not a viable gauge group for a GUT model, since it has only real representations. However, it contains as subgroups all the groups used for GUT models and a possible scenario is now that somehow quantum mechanics would compactify six more coordinates and at the same time break the E_8 group to a realistic GUT group.

v) String on orbifolds. The physically interesting models must have 4 large coordinates and the rest somehow compactified. For the strings so far discussed the 10-dimensional space M^{10} must split up into $M^4 \times K^6$: If we first regard K^6 as a hypertorus with all radii being the same we would satisfy the modular invariance [17], but the models will not be physically appealing since we will obtain an N=8 or N=4 supersymmetry which does not contain chiral fermions. Somehow we have to look for a less symmetric space. Such examples are provided by orbifolds. [18] Let us combine the six coordinates of K^6 to three complex ones, z_i, i = 1,2,3. Consider as one example the torus T made by the following identification of points (see Fig. 2):

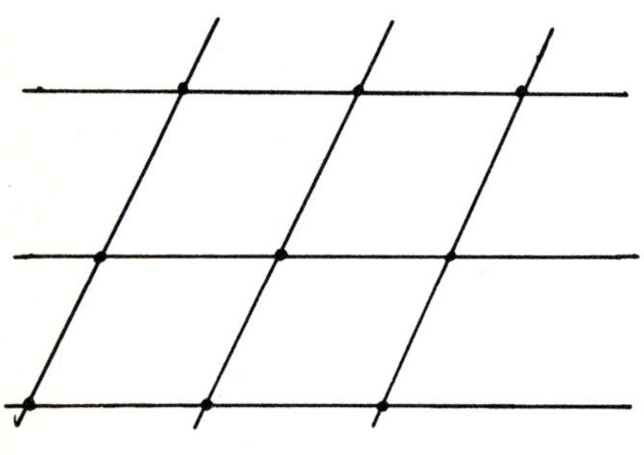

Fig. 2

$$z_i \approx z_i + 1 \approx z_i + e^{2i\pi/3} \quad .$$

This forms admits a Z_3 symmetry generated by

$$\alpha : z_i \to e^{2i\pi/3} z_i \quad . \tag{5.2}$$

On the torus there are 27 points left invariant by this transformation with

$$z_i = \frac{k}{\sqrt{3}} e^{i\pi/6} \qquad k = 0,1,2 \quad . \tag{5.3}$$

We would now like to include an additional equivalence relation that two points on T are equivalent if they are related by the α-action on the z_i:

$$\{z_i\} \approx \{e^{2i\pi/3} z_i\} \quad . \tag{5.4}$$

The coordinates z_i now span the space $z = T/\alpha$. This is not a manifold, because the equivalence relation introduces (conical) singularities at the 27 fixed points of α. Because of this it is called an orbifold. Even so we can let the string propagate on this manifold. The restriction is that we must project to α-invariant states by the projector

$$P = (1 + \alpha + \alpha^2)/_3 \quad . \tag{5.5}$$

Considering again the torus on which we check modular invariance we find that we sum over the boundary conditions in (σ,t)

$$(1,1) \quad , \quad (1,\alpha) \quad , \quad (1,\alpha^2) \quad .$$

This is obviously not modular invariant since we have already said that the first one is. In order to find a modular invariant expression we need to have different boundary conditions also in the σ-direction. In fact, we need to include all boundary conditions

$$(\alpha^k,\alpha^m) \qquad m,k = 0,1,2 \quad .$$

This amounts to introducing new sectors of particles, where $z(\sigma + \pi) = e^{ik\pi/3} z(\sigma)$, which we can check is all right also in the action since z is complex.

There are a number of reasons why string propagation on $M^4 \times Z$ is more realistic than on $M^4 \times T$. First of all compactification on $M^4 \times Z$ spontaneously breaks some of the supersymmetries. On an $SO(6) \approx SU(4)$ invariant torus there will be a 4 of supersymmetries. Now $\alpha \in SU(3)$ and to have α-invariance only one supersymmetry can remain unbroken, which is just what we want. This symmetry allows for chiral fermions and can also keep two scales in a GUT theory, i.e. a supersymmetry demands massless Higgs particles, which hence can be unaffected by the GUT scale.

This construction also allows for a breaking of the gauge group. Consider an element u in the gauge group which generates a Z_3 symmetry. In the above orbifold construction, instead of requiring that states be invariant under the rotation α, we could require them be invariant under the combined transformation

$$\bar{\alpha} = \alpha \cdot u \quad .$$

The whole construction above goes through. One now obtains a mechanism which also spontaneously breaks the gauge symmetry, namely the part of it which does not commute with u. In the case of the heterotic string with $E_8 \times E_8$ gauge group we can choose u to belong to the first E_8. This contains a maximal subgroup $E_6 \times SU(3)$. The center of SU(3) is Z_3 with a generator we can call u. The remaining symmetry group now becomes E_6 which is a considerable improvement, since E_6 is a viable GUT group.

This is just one of numerous examples of orbifold compactifications and there are several physically interesting ones.

vi) Strings on Calabi-Yau manifolds. The orbifolds considered above are in fact limiting cases of manifolds called Calabi-Yau manifolds [19]. We can obtain them from the orbifolds by cutting the canonical singularities and gluing back a smooth space. These manifolds were originally found as solutions to the low-energy interacting theory [20]. By assuming that the four-dimensional theory has an N=1 supersymmetry remaining one was led to these spaces. Although these are geometrically complicated spaces a lot of their topology can be deduced, which then can be used directly as information about the physical content. For example, the Euler characteristic tells how many generations the model carries and other topological invariants tell what the Yukawa couplings are.

One can also allow for a further breaking of the gauge symmetry down to the standard gauge group $SU(3) \times SU(2) \times U(1)$ times some U(1)'s [21]. This is obtained if the space contains further finite symmetries such that we can impose equivalence relations. Pictorially we can think of this as having holes in the space in which Wilson lines can be trapped. Only those gauge field which are not trapped survive. This effect is like a Higgs effect, but does not lead to the usual problem with Yukawa couplings that plague E_6 GUT-models.

There have been doubts raised that the Calabi-Yau spaces will survive in the full quantum theory. However, they represent very interesting examples where high-powered mathematics is very useful.

vii) Four-dimensional model. When we discussed the heterotic string we wanted a 10-dimensional model from the outset. It should, however, be clear that we could try to compactify further from the start. We could take 22 left-moving coordinates and 6 right-moving ones compactified on a torus [22]. This will introduce Lorentzian lattices $\Gamma_{22,6}$, where the two sets of coordinates have different metric (hence Lorentzian). Also here one can look for consistent solutions by checking the modular invariance. There are, in fact, a huge number of solutions!

In 2-dimensional theories such as theories on a world-sheet one can fairly easily make bosons out of fermions and vice versa. We can then bosonize the S^a_A's. The eight fermions are equivalent to four bosons and these can also be compactified as above, increasing the lattice even further and increasing the possible set of models also further.

It is clear at this stage that within the requirements imposed the set of possible models is huge, contrary to our belief that the set of possible string models should be quite limited. However, we must remember that all string models discussed above are constructed out of the same set of coordinates. They are different only because the boundary conditions on the coordinates are different.

It seems likely that all these models are consistent perturbatively. However, there are good reasons to believe that non-perturbative effects are very important and may be the tool to really constrain the set of possible models. We will discuss this fact in the next section.

6. LIGHT-CONE GAUGE FIELD THEORIES AND THEIR UNIQUENESS

The first-quantized algebra (3.10)-(3.13) can easily be rewritten in a second-quantized form. We first divide the phase space up into a set of coordinates mutually (anti-)commuting. Let us call a configuration of coordinates $\Sigma(\sigma)$. We then introduce an operator field $\Phi(\Sigma)$ with a commutation relation

$$[\Phi(\Sigma_1),\Phi(\Sigma_2)]_{x_1^+=x_2^+} = \frac{i}{2\partial^+_1}\Delta^{17}[\Sigma_1,\Sigma_2] \quad , \tag{6.1}$$

where Δ^{17} is a δ-functional which is only non-zero when the two configuration completely overlap.

The algebra (3.10)-(3.13) can now be written in terms of two-string functionals

$$G = -i \int D\Sigma\, \partial^{+}\Phi[\Sigma] g \Phi[\Sigma] \quad , \tag{6.2}$$

with g as in (3.10)-(3.13). In particular the Hamiltonian becomes

$$H = \int D\Sigma\, \partial^{+}\Phi[\Sigma] p^{-} \Phi[\Sigma] \tag{6.3}$$

and the free field equation is

$$[H,\Phi(\Sigma)] = -i\, \partial^{-}\Phi[\Sigma] \quad , \tag{6.4}$$

which can also be elevated to a Lagrangian and an action.

With this action, with appropriate boundary conditions for the coordinates imposed (which can be chosen according to the rules in the last section), we can ask if there is any freedom in it. For a point-particle theory describing gravity there is, since with a non-dimensionless Newton's constant κ one can have kinetic terms with an arbitrary power of the Klein-Gordon operator $\Box$.
This is a reflection of the fact that for Einstein's theory infinitely many new counterterms are possible. For the string theories this is not possible! Once the boundary conditions are chosen the kinetics are unique and quantum corrections cannot change this fact (unless there are anomalies).

It is also straightforward to introduce interactions in this scheme. We start by adding a three-string term to the Hamiltonian. In order to close the super-Poincaré algebra it is then necessary also to add three-string terms to all the generators that transform out of the quantization plane, namely Q_{-}, J^{+-} J^{i-}. This amounts to some guesswork again. We can start with the most general form for Q_{-} and H and start by demanding the commutator (3.9b) be true. The result of this analysis is once again unique [23]. For every model there is only one possible three-string interaction. There are no new counterterms that would spoil the quantum theory, which is the case for ordinary gravity.

It can also be shown that the actions are almost complete with just the three-string term. No four-string interaction, apart from a term which contributes just in the forward direction, is necessary [24]. Hence the structure of the actions is extremely restricted.

We have hence found that the freedom there is in string theory resides in the freedom to choose boundary conditions. At this stage the freedom is large and one might think that we are not at all on our way towards a unique model. However, we must remember that our understanding of the theories is still very limited. Especially we have no knowledge about the full quantum theories. The string concept will be important at Planck energies. There the effective coupling constant will be large (since Newton's constant is non-dimensionless). That means that the perturbation series is a series with even increasing terms. Can this sum be an asymptotic series, i.e. is the perturbation series really to be trusted? I think it is clear that more miracles must occur here. I agree with John Schwarz's words that "in string theories we expect miracles to happen", and I think that these miracles can very well select unique boundary conditions for the coordinates and hence a unique string model.

There are many aspects of string theory which I have not discussed here which also might constrain the possibilities. The perhaps most important problem is to find a geometric description of string theory. Important results have been presented on covariant field theories for bosonic strings and also for the superstrings. However, so far no geometric principle, such as the equivalence principle for Einstein's theory, has been presented. One problem here is that the principle behind string theory probably must be formulated for Planck-length physics where we have little intuition. When we discover this principle it will no doubt be of uttermost importance both for physics and mathematics. However, I think it will only be because of deep dynamical understanding that we might one day claim that we know the "final unified theory".

References

1. D.Z. Freedman, P. van Nieuwenhuizen and S. Ferrara, Phys. Rev. D13 3214 (1976)
2. P.S. Howe and U. Lindström, Nucl. Phys. B181 487 (1981)
   R.E. Kallosh, Phys. Lett. 99B 122 (1981)
   A.K.H. Bengtsson, L. Brink, M. Cederwall and M. Ögren, Nucl. Phys. B254 625 (1985)
3. For a review of string theory, see M.B. Green, J.H. Schwarz and E. Witten Superstring Theory I and II, CUP (1987)
4. P.A.M. Dirac, Rev. Mod. Phys. 26 392 (1949)
5. P. Goddard, J. Goldstone, C. Rebbi and C.B. Thorn, Nucl. Phys. B56 109 (1973)
6. L. Brink and H.B. Nielsen, Phys. Lett. 45B 332 (1973)
7. This method was suggested by R. Gliozzi, unpublished.

8. Y. Nambu, Lectures at Copenhagen Symposium, unpublished (1970)
T. Goto, Progr. Theor. Phys. 46 1560 (1971)

9. P.M. Ramond, Phys. Rev. D3 2415 (1971)
A. Neveu and J.H. Schwarz, Nucl. Phys. B31 86 (1971); Phys. Rev. D4 1109 (1971)

10. F. Gliozzi, D.I. Olive and J. Scherk, Phys. Lett. 65B 282 (1976); Nucl. Phys. B122 253 (1977)

11. M.B. Green and J.H. Schwarz, Nucl. Phys. B181 502 (1981)

12. M. Ademollo, L. Brink, A. D'Adda, R. D'Auria, E. Napolitano, S. Sciuto, E. Del Guidice, P. Di Vecchia, S. Ferrara, F. Gliozzi, R. Musto and R. Pettorino, Phys. Lett. 62B 105 (1976)

13. D.J. Gross, J.A. Harvey, E. Martinec and R. Rohm, Phys. Rev. Lett. 54 502 (1985); Nucl. Phys. B256 502 (1985); Nucl. Phys. B267 75 (1986)

14. L. Brink and H.B. Nielsen, Phys. Lett. 43B 319 (1973)

15. M.B. Green and J.H. Schwarz, Phys. Lett. 151B 21 (1985)

16. M.B. Green and J.H. Schwarz, Phys. Lett. 149B 117 (1984); Nucl. Phys. B255 93 (1985)

17. M.B. Green, J.H. Schwarz and L. Brink, Nucl. Phys. B198 474 (1982)

18. L. Dixon, J. Harvey, C. Vafa and E. Witten, Nucl. Phys. B261 678 (1985)

19. E. Calabi, A Symposium in Honor of S. Lefschetz. In Algebraic Geometry and Topology p. 78, Princeton Univ. Press (1955)
S.T. Yau, Proc. Natl. Acad. Sci. 74 1798 (1977)

20. P. Candelas, G. Horowitz, A. Strominger and E. Witten, Nucl. Phys. B258 45 (1985)

21. E. Witten, Nucl. Phys. B258 75 (1985)

22. H. Kawai, D.C. Lewellen and S.H. Tye, Phys. Rev. Lett. 57 1832 (1986)
I. Antoniadis, C. Bakhas and C. Kounnas, preprint A761 1286/LBL-22709
W. Lerche, D. Lüst and A.N. Schellekens, Nucl. Phys. B287 477 (1987)
W. Lerche, B.E.W. Nilsson and A.N. Schellekens, CERN-TH. 4692/87

23. A.K.H. Bengtsson, L. Brink, M. Cederwall and M. Ögren, Nucl. Phys. B254 625 (1985)

24 M.B. Green and J.H. Schwarz, Nucl. Phys. B243 475 (1984)
M.B. Green and N. Seiberg, Queen Mary preprint (1987)

String Theory and Quantum Gravity

T. Yoneya

Institute of Physics, University of Tokyo, Komaba, Tokyo, Japan

Recently there exists a great hope that the string theory may provide a completely unified theory of matter and forces including quantum gravity. In this talk, I would like to discuss the origins, meanings, and possible implications of the string theory as a finite theory of quantum gravity.

1. INTRODUCTION

One of the most crucial aspects of the string theory as a unification theory is that it includes the gravitational interaction in a quite natural and inevitable manner. Furthermore, there are several evidences for believing that the string theory can solve the divergence difficulty of the perturbative quantum gravity based on Einstein's General Relativity. However, our understanding of this problem seems to be still at a very infantile stage. For instance, even for the question why the string theory should include gravity, we do not have a completely satisfactory answer. This question must be a fundamental one in view of the fact that at sufficiently low energies the string theory gives the same scattering matrices in the tree approximation as those of Einstein theory. A satisfactory answer to the question should then explain how and in what sense the string theory is a generalization of Einstein (and super) gravity.

The purposes of my talk are first to summarize the relevant properties of the string theory related to these problems, and then to discuss the possible implications of these properties and of the crucial finiteness property of the string theory.

2. GRAVITY FROM STRINGS

Let us begin with a discussion about how the string theory contains gravity. For simplicity, the arguments will be restricted to bosonic string. First, I would like to remind you a well known fact as to why the graviton state, namely massless spin 2 state, is contained in the spectrum of string. This comes from the following properties:

(i) Unitarity requires that the Hilbert space of strings necessarily contains closed string states.
(ii) Spacetime Lorentz invariance.
(iii) World-sheet Weyl invariance.
(iv) World-sheet reparametrization invariance.

By (iii) and (iv), the equation of motion for the string coordinate $x^\mu(\tau,\sigma)$ ($\mu = 1,\cdots, D$) can always locally be written as $\partial_+\partial_- x^\mu = 0$ ($\tau^\pm = (\tau\pm\sigma)/\sqrt{2}$) with the constraints $(\partial_+ x)^2 = (\partial_- x)^2 = 0$. These equations are invariant under the conformal transformation $\tau^\pm \to f^\pm(\tau^\pm)$ which is the residual gauge symmetry of (iii) and (iv). Thus we find that the physical degrees of freedom are only D-2 out of D components for each right and left propagating modes $x^\mu_R(\tau-\sigma)$, $x^\mu_L(\tau+\sigma)$ of the string coordinate. Furthermore, the constraint $\int d\sigma\{(\partial_+ x)^2 - (\partial_- x)^2\} = 0$ demands that the world-sheet spatial momenta must always vanish and hence that the right and left modes must be excited simultaneously in closed strings. This means that the lowest excited level is represented as

Editors: N. Kawamoto and T. Kugo

$$\varepsilon^A_{\mu\nu}\alpha^\mu_{-1}\tilde{\alpha}^\nu_{-1}|0\rangle e^{ip\cdot x} , \qquad (A = 1,\cdots, (D-2)^2) \tag{1}$$

where α^μ_{-1} and $\tilde{\alpha}^\mu_{-1}$ are the creation operators corresponding to the wave modes $\exp -i(\tau-\sigma)$ and $\exp -i(\tau+\sigma)$, respectively, and p_μ is the spacetime D-momentum of the state

$$x^\mu(\tau,\sigma) = x^\mu + p^\mu\tau + i \sum_{n\neq 0}\frac{1}{n}(\alpha_n e^{-in(\tau-\sigma)} + \tilde{\alpha}_n e^{-in(\tau+\sigma)}).$$

How can the independent physical components of the state (1) be $(D-2)^2$ instead of D^2? As is well known, the constraint $(\partial_+ x)^2 = (\partial_- x)^2 = 0$ (the Virasoro condition) for the state (1) is represented as

$$\varepsilon^A_{\mu\nu}p^\mu = 0 = \varepsilon^A_{\mu\nu}p^\nu , \tag{2}$$

by which the D^2 components of (1) are reduced to $(D-1)^2$. Further reduction to $(D-2)^2$ in a compatible way with Lorentz invariance is possible only when $p^2 = 0$. The reason is that only in this case a state of the form $p_\mu\alpha^\mu_{-1}|0\rangle$ is a null state: namely, any state of the form $\varepsilon_\mu p_\nu\, \alpha^\mu_{-1}\tilde{\alpha}^\nu_{-1}|0\rangle$ or $p_\mu\varepsilon_\nu\alpha^\mu_{-1}\tilde{\alpha}^\nu_{-1}|0\rangle$ is of zero norm and orthogonal to the state (1) satisfying (2). Thus the polarization tensors $\varepsilon^A_{\mu\nu}$ and $\varepsilon^A_{\mu\nu} + \varepsilon^{(1)}_\mu p_\nu + \varepsilon^{(2)}_\nu p_\mu$ cannot be distinguished physically. In conformity with the symmetry argument, the "gauge transformation" $\varepsilon^{(1)}_\mu p_\nu + \varepsilon^{(2)}_\nu p_\mu$ can be interpreted as a conformal transformation,

$$(\varepsilon^{(1)}_\mu p_\nu + \varepsilon^{(2)}_\nu p_\mu)\alpha^\mu_{-1}\tilde{\alpha}^\nu_{-1}|0\rangle = (L_{-1}\varepsilon^{(1)}\cdot\tilde{\alpha}_{-1} + \tilde{L}_{-1}\varepsilon^{(2)}\cdot\alpha_{-1})|0\rangle, \tag{3}$$

where L_n, $\tilde{L}_n$ are usual Virasoro operators

$$L_n = \frac{1}{4\pi}\int_0^{2\pi} d\sigma(\partial_+ x)^2 e^{in\sigma} ,$$
$$\tilde{L}_n = \frac{1}{4\pi}\int_0^{2\pi} d\sigma(\partial_- x)^2 e^{in\sigma} . \tag{4}$$

It is well known that any state of the form $\sum_{n=1}^\infty\{L_{-n}|\Lambda_n\rangle + \tilde{L}_{-n}|\tilde{\Lambda}_n\rangle\}$ decouples from the physical on-shell scattering amplitudes of strings.

The conclusion drawn from this argument is that if the closed string sector is correctly quantized in a compatible manner with Weyl, reparametrization and Lorentz invariance, the first excited state of closed bosonic string must necessarily be massless and has $(D-2)^2$ independent physical components. In terms of the irreducible representations of Lorentz group, there are three different representations: the traceless symmetric components $\frac{1}{2}D(D-3)$ (graviton), the antisymmetric components $\frac{1}{2}(D-2)(D-3)$, and a trace component (dilaton). Combining with the modular invariance argument of vacuum energy [1] which demands that the Regge intercept must be $(D-2)/12$, the above conclusion also leads to the critical spacetime dimension $D = 26$.

The general unitarity (and Lorentz invariance) of the on-shell S-matrices with respect to $(D-2)\times(D-2)$ coordinate modes, of which the $D(D-3)/2$-component "graviton" state is a special case, is guaranteed by the built-in gauge invariance reflecting the freedom of world-sheet conformal transformations.

From the spacetime point of view, however, the appearance of "graviton" in string theory seems rather mysterious. In the case of Einstein theory, the graviton is the quantum associated with the fluctuation of the spacetime metric tensor $g_{\mu\nu}(x)$ which has $D(D+1)/2$ components. The reduction of $D(D+1)/2$ to $D(D-3)/2$ occurs as follows. The general coordinate invariance $x^\mu \to x^\mu + f^\mu(x)$ not only reduces the meaningful components of $g_{\mu\nu}$ to $D(D+1)/2 - D = D(D-1)/2$, but also implies that the D "time"-components $R_{\mu 0} = 0$ out of D(D+1)/2(-D) components of the field equation $R_{\mu\nu} = 0$ ($\nabla^\mu(R_{\mu\nu} - \frac{1}{2}g_{\mu\nu}R) = 0$) for the on-shell graviton are actually the constraints on the Cauchy data in the same sense that the Gauss law is the constraint in Maxwell's electrodynamics. Consequently, the physical components of $g_{\mu\nu}$ is D(D-1)/2 - D = D(D-3)/2. Thus, the unitarity of the theory is in this case a consequence of spacetime general coordinate invariance. The string theory does not introduce the spacetime metric as the dynamical variable, nor presupposes the fundamental requirements of General Relativity, namely the principle of equivalence and the spacetime general coordinate invariance. Nevertheless, the symmetries of the world-sheet quantum dynamics of string reproduce a similar consequence as General Relativity.

There are evidences showing that this is not a mere acciedent. Firstly, it has been well known from explicit computations and general arguments [2,3] that the on-shell tree scattering amplitudes of the string graviton reduce to those of Einstein theory in the limit of infinite string tension (zero-slope limit). This amounts to saying that the effective action for the interactions of string graviton in the tree approximation is nothing but the Einstein-Hilbert action in the low energy limit. Secondly, the requirement of consistent propagation of strings preserving the properties (i)~(iv) in a curved background spacetime demands that the background spacetime metric satisfies certain "fixed-point condition" of the renormalization group in the sense of two-dimensional field theories on the string world sheet. In the zero-slope limit, the fixed point condition coincides [4] with the Einstein equation at critical spacetime dimensions. Thus the same dynamics, Einstein theory, governs both the spacetime background and the fluctuations upon it at sufficiently large distances in string theory.

3. "EQUIVALENCE PRINCIPLE" AND "GENERAL COVARIANCE" IN STRING THEORY

At present, we have to admit that the string theory is essentially a collection of various game of rules for constructing perturbatively the on-shell S-matrices corresponding to the spectra of quantum free strings. The basic problem is therefore to find a non-perturbative framework of string theory from which the known perturbative rules can be derived and the fundamental questions, such as the vacuum structure and spacetime compactification, are meaningfully posed. The deeper understanding of the intriguing connection between string theory and General Relativity (and/or Supergravity) would perhaps be useful in that direction. From this point of view, it seems worthwhile to ask whether and how, if any, the equivalence principle and/or the spacetime general covariance are embodied in the present formalism of string theory.

As is well known, there is a one-to-one correspondence between the physical free string states and the physical vertex operators, i.e., world-sheet *local* operators of conformal dimension 2. This means that any on-shell string state can be regarded as an allowable external field by adding the corresponding vertex operator to the world sheet action and vice verse:

$$S_{\text{free string}} + \sum_n \phi_n V_n = S_{\text{string in external field}}, \tag{5}$$

where ϕ_n's are the spacetime fields corresponding to the vertex operators V_n's. It is a remarkable property that the interaction of extended objects can be equivalently represented by the presence of *local* external fields. In particular, for massless on-shell physical states such as the string graviton no *ad hoc* normal-ordering prescription is necessary for defining the vertex operator. This explains why the string "graviton" behaves as the graviton even in interactions, not merely as an accidental massless spin-2 particle in the free string spectra. Namely, the vertex operator for the string graviton is the transverse part of the *spacetime* energy-momentum tensor of the string

$$V^{\mu\nu}(k) = \int d^2\xi \, e^{ikx(\xi)} \partial x^\mu(\xi) \cdot \partial x^\nu(\xi), \tag{6}$$

which is by definition the variation of the world-sheet action with respect to the external spacetime metric $g_{\mu\nu}$. This result is valid for any string models including spinning strings [5]. Thus we can say that the gravitational interaction of string models satisfies the principle of equivalence in the limit of weak interaction. Furthermore, as a consequence of (6), the string theory automatically satisfies the spacetime general coordinate invariance in the following sense. Consider the effect of an infinitesimal spacetime coordinate transformation $x^\mu \to x^\mu + f(x^\mu)$ satisfying $\Box f^\mu = 0 = \partial^\mu f_\mu$ on a one-string state $|\psi\rangle$ corresponding to a Riemann surface with a boundary. The Euclidean path-integral representation of the change of the single-string state $|\psi\rangle$ under the coordinate transformation is essentially given by

$$\delta_f|\psi\rangle = \int_{x=y \text{ on } \partial M} \frac{[d^Dx][dg]}{V_{local}} e^{-S_{free}} \int d^Dk\delta(k^2)V^{\mu\nu}(k)\{\tilde{f}_\mu(k)k_\nu + \tilde{f}_\nu(k)k_\mu\}, \quad (7)$$

where the path integral is performed with an appropriate boundary condition on an arbitrary Riemann surface M with a boundary ∂M, and $f(x) = \int d^Dk\delta(k^2)\tilde{f}(k)\exp ikx$. The restriction $\Box f^\mu = 0 = \partial^\mu f_\mu$ is necessary for the well definedness of the vertex operator and the transformation of the path-integral measure $[d^Dx]$ under the coordinate transformation $x^\mu \to x^\mu + f^\mu(x)$. Equivalently, (7) can also be represented as

$$\delta_f|\psi\rangle = i\int_0^{2\pi} d\sigma\; f^\mu(x(\sigma))p_\mu(\sigma)\;|\psi\rangle \quad (8)$$

under the restriction for f^μ, where $p_\mu(\sigma)$ is the conjugate momentum density of the string coordinate at ∂M. We know from the discussion of the previous section that the state (7) decouples from the physical S-matrices. This means that in string theory the spacetime general coordinate invariance in certain *restricted* sense is automatically satisfied without introducing independent spacetime metric field.

Evidently a crucial aspect of string theory ensuring these remarkable roles of the string graviton is the equivalence between the string spectrum and the vertex operators. The fact that it naturally explains how the fundamental principles of General Relativity are partially built-in in string theory suggests something deeper behind, despite its apparent triviality. At present, however, it seems that we do not know appropriate languages for expressing the "something".

One of the interesting properties related to this observation is the conjectured relationship [6] between the effective action $\Gamma(\phi)$ for the S-matrices and the β-function of the renormalization group approach:

$$\beta_n(\phi) = \sum_m G_{nm}(\phi)\frac{\delta\Gamma(\phi)}{\delta\phi_n}\;, \quad (9)$$

where $G_{nm}(\phi)$ is an appropriate metric in the space of vertex operators. (9) has been confirmed to a few lower orders both with respect to α'- and ϕ- expansions [7]. If (9) is satisfied, $-\Gamma(\phi)$ may be identified with the Zamolodchikov "C-function" ($\Sigma\beta_n(\phi)\frac{\delta}{\delta\phi_n}C(\phi) < 0$) [8], which then gives a simple general expression for the effective action in terms of the 2-dimensional (world-sheet) energy-momentum tensor. This will perhaps be useful for clarifing the relation between the spacetime and world-sheet symmetry properties.

It is also tempting to speculate the speciality of the string model discriminating from other extended objects on the basis of the state-vertex correspondence. The fundamental reason why the one-shell string state can be described by the local vertex operators is the conformal equivalence between an infinitely long tube in a world sheet and a node made by pinching off the surface around a non-trivial cycle.

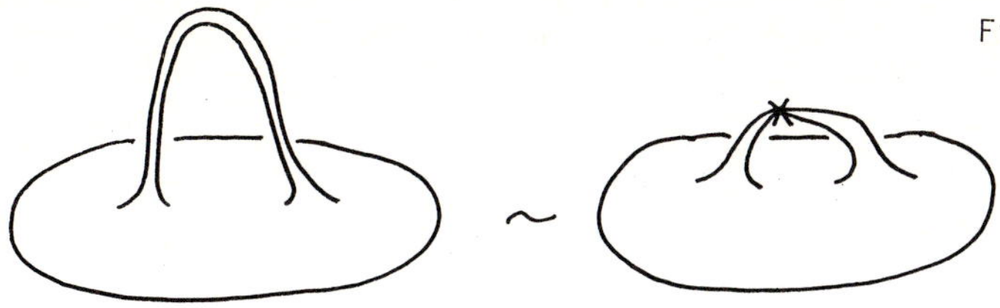

Fig.1

Thus the interaction among on-shell states can be described by insertion of local vertex operators on a Riemann surface. This conformal equivalence is of course a direct consequence of the properties (iii) and (iv). If instead of string some higher dimensional extended object is adopted as a model for unification, a crucial necessary (not sufficient) condition for the model to naturally describe the gravitational interaction *without* independent metric field must thus be the Weyl invariance. Unless the Weyl invariance is maintained, the massless spin-2 state, even if it occurred in free spectrum, could not replace the role of the fluctuation of the spacetime metric. It is, however, very difficult to find Weyl invariant higher dimensional models which are *viable* for quantization. Also it seems unclear whether there is any natural mechanism ensuring the emergence of massless spectrum in such higher dimensional models.

4. COMMENT ON THE NATURE OF STRING FIELDS

One of the presently familiar attempts at non-perturbative formulation of string theory is the covariant string field theory. Since this approach clarified the built-in gauge structure of the string theory associated with conformal transformation and furthermore suggested a possibility of background independent reformulation, it is appropriate here to briefly review its achievements and problems.

The essence of covariant string field theory is to regard the BRST-cohomological structure used in the characterization of physical space [9] as a consequence of a generalized spacetime gauge structure associated with the string field. A string field A is assumed to transform as $A \to A + Q_B\Lambda + g\Lambda \times \Lambda + \cdots$, a generalization of the cohomological transformation $|\psi\rangle \to |\psi\rangle + Q_B|\Lambda\rangle$ of the free-string wave function $|\psi\rangle$, where the notation $\times$ is an appropriate rule for forming the product of string fields, and Q_B is the BRST-charge operator. Indeed, the linearized approximation of the BRST-gauge transformation reproduces the linearized forms of the usual gauge transformations of local gauge fields including the case (3) of gravitational fields. At present, the most successful construction of covariant string field theory seems to be the one due to Witten [10] for open strings. There are also a series of works done by Kyoto and CERN groups [11]. Both of these results will be discussed by other speakers in this symposium. So I will only point out some basic open problems putting aside the technical aspects.

My basic question is whether the covariant string field as presently formulated in the BRST formalism is the genuine dynamical degree of freedom which should be subject to the ordinary rules of quantization. In covariant open string field theories such as Witten's one, the closed string only appears [12] as a loop effect. This raises the following questions. Firstly, the *perturbative* unitarity requires that the closed string state is actually contained in the Hilbert space of multi-open strings. Remember an apparently similar phenomenon in the 2 dim. Schwinger model: The one-loop vacuum polarization diagram generates a scalar pole, which is in this case consistent with perturbative unitarity since the scalar pole is due to the propagation of massless collinear fermions.
In the case of Witten's string field theory, it is not clear how similar things could happen because in this case the pole singularity is nothing to do with the Landau singularity induced by the propagation of open strings. Secondly, it is difficult to understand how the purely gravitational (i.e. closed string) contribution to the vacuum energy arises from open string field theory whose dynamical degree is only the open string field. Since the closed string graviton

is contained in the theory, we usually expect the vacuum energy (= cosmological constant) should receive the contribution also from the graviton loop. Note that the Euler numbers of the closed string and open string one loops are both zero. Of course, these problems (especially the second one) do *not* disprove anything. But it is certainly true that there are several points to be clarified further.

For closed string theories, it is fair to say that at present we have no satisfactory construction at the same level as Witten's theory for open strings. There is the same basic question about the nature of covariant closed string field if it can be formulated. If the covariant string field is subject to the usual rules of covariant quantization, there should be a natural triangulation of the moduli space of closed Riemann surfaces corresponding to the usual Feynman rules. Even in the tree approximation, there is no such natural triangulation, except for the light-cone diagrams [13] based on the Mandelstam mapping. It should also be pointed out that the familiar prescription of the fundamental integration region for the vacuum diagrams of closed strings cannot be understood in the framework of string field theories. This problem is present already at the free level *if* the covariant string field is just a gauge field with infinite number of components as usually assumed [14].

The followings might be the possible resolutions of the questions.

(1) Ordinary covariant gauge fixing (such as the Siegel gauge) may not be allowed because of the Gribov-like problem.

(2) Find larger gauge symmetry structure which allows the light-cone gauge fixing *off-shell*. Or more drastically,

(3) change the view point. Namely, the concept of string field in the sense of the BRST formalism is not the primordial dynamical entity which is subject to the usual rules of quantization and plays only a secondary role in connecting the fundamental non-perturbative formulations of string theory and the S-matrices. If this view is correct, what could be the primordial entity? In any case at the present stage of developments, it seems advantageous to accept various possibilities by not sticking on a particular approach.

To summarize what has been discussed so far. We see that the string theory may certainly be a profound generalization of Einstein's General Relativity (and Supergravity). But we do not have appropriate languages for expressing the full non-perturbative content of the string theory.

Of course, the string theory receives revived interests recently not just because it is a miraculous generalization of General Relativity, but because physicists began to recognize the possibility of uniting quantum theory and gravity through strings, since the latter may solve the problems of divergences and anomalies. In the rest of this talk, I would like to discuss possible implications concerning about the finiteness of string theory.

5. FATE OF SPACETIME CONTINUUM?

It is often said that the slope parameter α' sets a natural cutoff momentum of order $\sqrt{\alpha'}^{-1}$ ($\sim$ Planck energy) due to the extended nature of strings. Although this is basically correct, it is not entirely obvious whether interacting models of continuously extended objects are always ultraviolet finite, since the objects themselves contain additional infinite number of degrees of freedom. The latter may give rise to new divergences. In fact, this would actually be the case when the extended object is 3 or higher dimensional, because then they cannot be described, in parametrization independent way, by a renormalizable field theory.

As is now well known, the key mechanism of (ultraviolet) finiteness of (closed) string theory is the modular invariance [1]. The modular invariance is the special case of the requirements (iii) and (iv). Specifically, it is derived from the reparametrizations which are homotopically inequivalent to the identity transformation and from the associated Weyl rescalings. The interchange of the "time-like" variable τ (= Schwinger's proper time) and the "space-like" variable σ (= parametrization of a string) is a typical example. It should be stressed that the association of Weyl rescaling is crucial to the modular invariance. These transformations are a generalization of the duality transformation of string tree diagrams. In this sense, the modular invariance is a natural mathematical representation of the notion of duality, and hence duality is responsible for the ultraviolet finiteness of the string theory. Duality, i.e., modular invariance, ensures that there is no ultraviolet singularity in the string diagrams. The unitarity singularities (which are infrared) are smoothly connected to each other without encountering the ultraviolet singularities, in sharp contrast with the situation in the usual point-particle Feynman diagrams.

Now I would like to propose to ask what is the proper formulation of this remarkable property in terms of *spacetime* languages. We have learnt that the string theory automatically contains a graviton which can be interpreted as representing the fluctuation of the spacetime geometry in the restricted sense ($\alpha' \neq 0$) or in the zero-slop limit ($\alpha' \to 0$). This implies that the very notion of string itself must be a more fundamental geometrical entity encompassing the notions of metric tensor, connection, curvature and so forth, of Riemannian geometry which is the appropriate language for General Relativity based on local field theories. Note that the string theory by definition assumes that everything is made out of strings. Even the geometry of spacetime must be expressed by using only the notion of strings. This seems require a fundamental revision on the usual concept of the spacetime continuum. Then the constant α' should be interpreted as putting certain limitation on the usual concept of Riemannian spacetime, just as the Planck constant h put a limitation on the classical concept of the phase space. From this point of view, we expect some clear characteristic relation expressing the limitation, in analogy with the commutation relation or Heisenberg's uncertainty principle in quantum theory.

I now would like to suggest a possible hint along this line of arguments. Namely, a possible expression of duality in the spacetime languages could be

$$\Delta t \cdot \Delta \ell \gtrsim 4\pi\alpha' , \tag{10}$$

where Δt is an appropriate measure of the indeterminacy of the propagation length of a string state and the $\Delta \ell$ is an appropriate measure of the indeterminacy of the intrinsic extendedness of a propagating string. This "spacetime indeterminacy relation" means that any observation in string theory probing short "time" ($\Delta t \to 0$) or small "distance" ($\Delta \ell \to 0$) structure in spacetime is associated with large indeterminacy in the "dual" variable, $\Delta \ell$ or Δt, respectively. Thus (10) sets a limitation about the smallness of the spacetime domain where arbitrary possible observation in string theory is performed.

In the limit of small Δt, (10) can in fact be derived [15] by reinterpreting the Heisenberg relation $\Delta t \Delta E \gtrsim h$. Here instead of doing that, let us first consider the behavior of the string propagator in the light-cone gauge. ($\lambda^2 = 4\pi\alpha'$)

$$\Delta[x(\sigma), y(\sigma)] = \int_0^\infty \frac{d\tau}{\tau^{13}} e^{\frac{4\tau}{\lambda^2}} \prod^{\infty} (1 - e^{-\frac{4n\tau}{\lambda^2}})^{-24} \exp - S[x,y] , \tag{11}$$

$$S[x,y] = - \frac{1}{2\tau}(x_0 - y_0)^2 + \frac{\pi}{\lambda^2} \sum_{n=1}^{\infty} \frac{1}{\sinh\frac{2n\tau}{\lambda^2}} [(|\mathbf{x}_n|^2 + |\mathbf{y}_n|^2)\cosh\frac{2n\tau}{\lambda^2} - 2\mathrm{Re}(\mathbf{x}_n \cdot \mathbf{y}_{-n})] , \tag{12}$$

$$x^i(\sigma) = \frac{1}{\sqrt{\pi}} x_0{}^i + \sum_{n=1}^{\infty} x_n{}^i e^{-in\sigma}$$

$$y^i(\sigma) = \frac{1}{\sqrt{\pi}} y_0{}^i + \sum_{n=1}^{\infty} y_n{}^i e^{-in\sigma} . \tag{13}$$

When $(x_0-y_0)^2$ is finite, the dominant integration region for τ satisfies $\tau \gtrsim (x_0 - y_0)^2$. The smaller $\Delta t \sim \sqrt{(x_0-y_0)^2}$ becomes, the region extends to the smaller τ. In the limit of small τ, the second term of (12) behaves as

$$\left(\frac{1}{2n\tau} - \frac{n\tau}{3\lambda^4} \right) |\mathbf{x}_n - \mathbf{y}_n|^2 + \frac{n\tau}{\lambda^4}(|\mathbf{x}_n|^2 + |\mathbf{y}_n|^2) + O((n\tau)^3) \tag{14}$$

for n satisfying $\frac{2n\tau}{\lambda^2} \ll 1$. (14) implies that the intrinsic extension of the propagating state is characterized by the quantity $\lambda^2/\sqrt{\tau}$. Hence,

$$\Delta t \cdot \Delta \ell \sim \lambda^2 = 4\pi\alpha' \, . \tag{15}$$

On the other hand for sufficiently large $(x_0-y_0)^2$, the contribution to (11) comes from the region of large τ, where the second term (12) behaves as

$$\frac{1}{\lambda^2} \sum_{n=1} n(|\mathbf{x}_n|^2 + |\mathbf{y}_n|^2) \, . \tag{16}$$

(16) is just the exponent of the wave function of a free on-shell string as it should be. This of course does not contradict with (10). However, the vertex-state correspondence, the "equivalence principle" of string theory, says that the massless on-shell string states can be treated as if they are *local* external fields. In this sense, effectively $\Delta\ell \to 0$ for such states. (10) would then require $\Delta t \to \infty$.

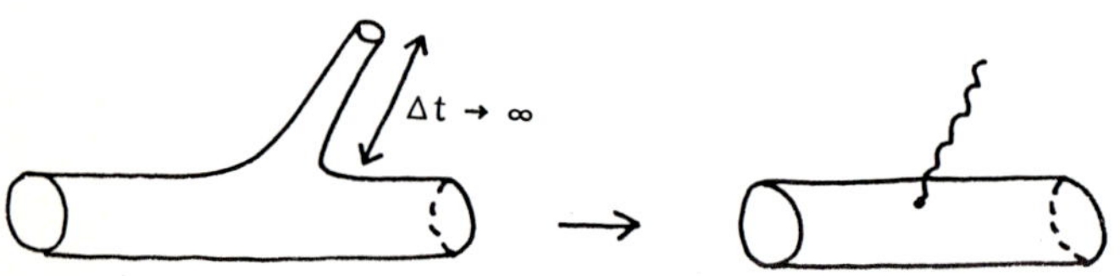

Fig.2

(10) also conforms to the correspondence between the non-planar orientable one-loop diagram and the closed-string tree diagram. The annulus diagram, with the inner (outer) boundary being of radius r (1), of the former picture is characterized by the interaction time $\Delta t \sim \lambda(\log r^{-1})^{-1}$ as is easily convinced by inspecting the usual trace calculation. (In the familiar notation [16], $(\log r^{-1})^{-1} \sim \log w$.)

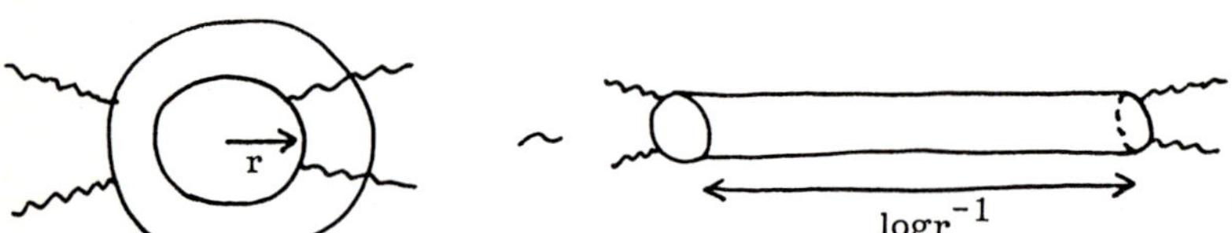

Fig.3

The annulus diagram is reinterpreted by the Jacobi imaginary transformation as a closed string diagram in which the proper time of the propagation of closed string is of the order $\lambda(\log w)^{-1}$. In this case t and the proper time of the closed string are mutually dual. Hence $\Delta t \cdot \Delta\ell \sim \lambda^2$. This is certainly a consequence of duality. It is to be noticed that the Jacobi imaginary transformation of annulus is the counterpart of the modular transformation in the torus diagram.

Admittedly, these arguments are not yet sufficiently precise. In particular, we have not given mathematical definitions for the measures Δt and $\Delta\ell$. In order for the relation (10) to reflect the "equivalence principle" in a more precise way, there should be a conformally invariant way of formulating it.

As is well known, the Planck length $\ell_{p\ell} \sim \sqrt{Gh/c^3}$ can be interpreted as characterizing a limiting length for the simultaneous application of General Relativity and Quantum Mechanics. In the string theory for unification, the role of Planck length is replaced by the slope parameter α'. To conclude, I would like to stress that a clarification of the nature of the limitation set by α' may possibly be one of the important aspects for future development of string geometry.

This work is partially supported by Grant-in-Aid, the Ministry of Education, Science and Culture of Japan (No. 60540170).

References

1. J.A. Shapiro: Phys. Rev. D5, 1945 (1972). See also L. Brink and H.B. Nielsen: Phys. Lett. 45B, 332 (1973).
2. T. Yoneya: Prog. Theor. Phys. 51, 1907 (1974); Lett. Nuovo Cim. 8, 951 (1973).
3. J. Scherk and J.H. Schwarz: Nucl. Phys. B81, 118 (1974); Phys. Lett. 52B, 347 (1974).
4. D.H. Friedan: Phys. Rev. Lett. 51, 334 (1980); Ann. Phys. (NY) 163, 318 (1985).
5. T. Yoneya: Nuovo Cim. 27A, 440 (1975); Prog. Theor. Phys. 56, 1310 (1976).
6. A.M. Polyakov: SLAC preprint Trans-0222 (1986). See also, C. Lovelace: Nucl. Phys. B273, 413 (1986).
7. C.G. Callan, D. Friedan, E.J. Martinec and M.J. Perry: Nucl. Phys. B262, 593 (1985). For recent developments, see, for example, A.A. Tseytlin: Phys. Lett. 194B, 63 (1987) and references therein.
8. A.B. Zamolodchikov: JETP Lett. 43, 730 (1986).
9. M. Kato and K. Ogawa: Nucl. Phys. B212, 443 (1983).
10. E. Witten: Nucl. Phys. B268, 253 (1986).
11. H. Hata, K. Itoh, T. Kugo, H. Kunitomo, and K. Ogawa: Phys. Rev. D34, 2360 (1986); A. Neveu and P. West: Nucl. Phys. B278, 601 (1986).
12. S.B. Giddings, E. Martinec and E. Witten: Phys. Lett. 176B, 362 (1986).
13. M. Kaku and K. Kikkawa: Phys. Rev. D10, 1823 (1974).
14. For further discussions on the nature of string fields, see T. Yoneya: In *Seventh Workshop on Grand Unification* /ICOBAN '86, ed. by J. Arafune (World Scientific, 1987).
15. T. Yoneya: In the volume commemorating the 60-th birthday of Professor Kazuhiko Nishijima (World Scientific, to be published).
16. See, e. g., J. H. Schwarz: Phys. Rep. 89, 223 (1982).

Part II

String Field Theories

Construction of Witten's String Field Theory

A. Jevicki

Department of Physics, Brown University, Providence, RI02912, USA

I. Introduction

During the last several years there has been exciting progress in the construction of field theory of strings. Originally, it was in the light cone gauge [1-5] that the theory was constructed first and the challenging task was the one of giving a covariant and gauge invariant theory [6-14]. This was successfully achieved for open strings with some interesting ideas for closed strings. The theory outlined by Witten [15,16,17] will be the topic of this review. We shall describe its complete operator construction and emphasize its successes. Most of the review (Section II and Section III) is based on the work [17,18] done in collaboration with D. Gross. In the last section we discuss some more–recent developments and open questions.

One of the main goals in developing string field theory (apart from its physical usefulness) is in the possibility that it will provide a novel and deeper insight into the basic structure of string theory. The open string theory contains an interesting mathematical structure. However, as yet there has not been a fundamental breakthrough and the theory of closed strings has eluded construction. Pursuing this direction, however, is rather relevant; it might, in principle, lead to a more general framework going even beyond field theory itself.

II. Basics of String Field Theory

In the covariant approach to string field theory the string field A is a functional of the basic string variables. These in the orthonormal gauge are the coordinates $X_\mu(\sigma,\tau)$, $\mu = 1\dots 26$ and also the ghost and antighost variables $c(\sigma,\tau)$ and $b(\sigma,\tau)$. The fermionic ghost variables can also be replaced by a bosonized ghost $\phi(\sigma)$. The string field obeys the following hermiticity condition

$$A^+[x(\sigma),c(\sigma)] = A[x(\pi-\sigma),-c(\pi-\sigma)] \; . \tag{1}$$

The ghost number operator is

$$N_{gh} = \frac{1}{2}(c_0b_0 - b_0c_0) + \sum_{n>0}(c_{-n}b_n - b_{-n}c_n) \tag{2}$$

and the string field has ghost number $-\frac{1}{2}$. That is the ghost number associated with the $|b_0 = 0>$ state. The free action is simply

$$S_0 = (A, QA) \; . \tag{3}$$

Here a fundamental role is played by the BRST operator

$$Q = \int d\sigma : C(\sigma)(L^x(\sigma) + \frac{1}{2}L^{gh}(\sigma)) : \tag{4}$$

Springer Proceedings in Physics Vol. 31: **Quantum String Theory**
Editors: N. Kawamoto and T. Kugo

The nilpotency of Q ($Q^2 = 0$) implies the gauge symmetry

$$A \to A' = A + Q\Lambda \ . \tag{5}$$

The gauge fixing procedure for the free action is slightly nonstandard as it involves the phenomena of ghosts.

The basic starting point of Witten's interacting theory [15] is a differential calculus in string space. Namely the nilpotency of Q offers an interpretation of this operator as a derivation (in string space). The next operator to introduce is then an integration $\int$ and finally a wedge product $*$. One can use a bosonized language for the ghosts where they are replaced by a scalar field $\phi(\sigma)$ (we shall often treat it as a 27-th coordinate). The measure for the definition of integration introduced by Witten reads

$$\int A = \int \mathcal{D}X_\mu \mathcal{D}\phi e^{-3i\phi(\frac{\pi}{2})/2} \cdot \prod_{0 \leq \sigma \leq \pi/2} \delta\left[X_\mu(\sigma) - X_\mu(\pi - \sigma)\right] \cdot$$

$$\cdot\, \delta\Big(\phi(\sigma) - \phi(\pi - \sigma)\Big) A[X_\mu, \phi]. \tag{6}$$

It represents a folding of the string about its midpoint with the phase factor supplying the ghost number. The basic reason for the above definition is to achieve the relevant property of integration, that is

$$\int QA = 0 \ . \tag{7}$$

The next operation is the wedge product $*$: with two string functionals A and B one associates a third one $C = A * B$. The procedure is that of annihilating half of string one with a half of string 2 with the remaining halves rearranged into string 3. This algebra is not commutative $A * B \neq B * A$ but it is associative

$$A * (B * C) = (A * B) * C \ . \tag{8}$$

Moreover Q is to act as a derivative on the $*$

$$Q(A * B) = (QA) * B + (-1)^A A * (QB) \tag{9}$$

with $(-)^A = (-)^{N_{gh} - \frac{1}{2}}$. Further axioms are

$$\int A * B = (-)^{AB} \int B * A \tag{10}$$

$$(-)^{QA} = -(-)^A \ . \tag{11}$$

Witten gave arguments for Q or BRST invariance based on the current conservation on analogue Riemannian surfaces. The ghost number assignments follow from the Rieman-Roch theorem applied on the analogue surfaces. Likewise BRST invariance follows from a conservation of the BRST current.

In the language outlined the gauge invariant free string action does have a natural representation, namely

$$S_o = \int A * QA \ . \tag{12}$$

The $*$ operation then provides a three–string interaction term $\int A * A * A$. This corresponds to the three–string overlap interaction

$$V_3 = e^{3i\phi(\frac{\pi}{2})/2} \prod_{0\leq\sigma\leq\pi/2} \prod_{r=1}^{3} \delta\left[X^r(\sigma) - X^{r-1}(\pi-\sigma)\right] \ , \tag{13}$$

where the phase factor specifies the ghost number $\frac{3}{2}$. These two terms represent the whole string field theory Lagrangian

$$S = \int (A * QA + \frac{2}{3} A * A * A) \tag{14}$$

with the gauge invariance

$$\delta A = Q\Lambda - \Lambda * A + A * \Lambda \ . \tag{15}$$

This gauge invariance is a basic consequence of the axioms obeyed by the $Q, *, \int$ system. In addition to gauge invariance there also exists a special global symmetry in this theory. The following subgroup

$$K_n = L_n - (-)^n \ L_{-n} \tag{16}$$

of the Virasoro algebra leaves all the δ-function overlaps invariant. These generators induce reparametrization transformations which are odd under $\sigma, \tau \to \pi - \sigma, -\tau$ and they naively leave the midpoint invariant. A special characteristic of this subalgebra (of the Virasoro algebra) lies in the fact that it is free of central charge anomalies. Namely both the orbital K_n^x and the ghost generators K_n^{gh} obey

$$\left[K_n, K_m\right] = (n-m) \ K_{n+m} - (-)^m (n_m) K_{n-m} \ . \tag{17}$$

At the quantum level all these transformations can and often do have midpoint anomalies which in general could endanger the symmetries. Indeed, it was found that the vertex operators are not annihilated by the generators K_n^x or K_n^{gh} but instead

$$K_{2N} = CN(-)^N \ , \tag{18}$$

where C are constants. Moreover, it is also found that for the total generator $K_n = K_n^x + K_n^{gh}$ these anomalies cancel (in $D = 26$) leading to a symmetry [17].

We now begin with the outline of an operator formulation of the theory. This approach consists in representing all the operations involved in the Hilbert space of string creation-annnihilation operators. This formulation gives a precise definition of the theory and most importantly provides a concrete computational framework.

The δ-function overlaps define the equations which are to be solved for the integration, $*$ operation and the interaction vertex. Consider, for example, integration of a string functional. For the moment we ignore the ghost degrees of freedom and represent the integral as

$$\int A \equiv \int \mathcal{D}x(\sigma) \prod_{0\leq\sigma\leq\frac{\pi}{2}} \delta[x(\sigma) - x(\pi-\sigma)] A[x(\sigma)] \equiv \langle I \mid A \rangle \ , \tag{19}$$

where $\mid A\rangle$ is the vector in Fock space corresponding to the string functional field $A\left[x(\sigma)\right]$

$$A[x(\sigma)] = \langle x(\sigma) \mid A\rangle \tag{20}$$

and $\mid I\rangle$ represents integration:

$$I[x(\sigma)] = \langle x(\sigma) \mid I\rangle = \prod_{0\leq\sigma\leq\frac{\pi}{2}} \delta\big(x(\sigma) - x(\pi - \sigma)\big)\ . \tag{21}$$

I is the identity functional, with respect to $*$. Its precise meaning is given by constructing explicitly the vector $\mid I\rangle$. With the normal mode expansions,

$$x(\sigma) = x_o + \sqrt{2}\sum_{n=1}^{\infty} x_n \cos\, n\sigma, \tag{22}$$

we have the creation and annihilation operators:

$$x_n = \frac{i}{2}\sqrt{\frac{2}{n}}(a_n - a_n^\dagger)\ . \tag{23}$$

The overlap equations for I

$$(x(\sigma) - x(\pi - \sigma)) \mid I\rangle = 0 \tag{24}$$

can then be written out in terms of the oscillators. The equation is readily solved and gives

$$\mid I\rangle = e^{-\frac{1}{2}a_{-n}(-)^n a_{-m}} \mid 0\rangle\ . \tag{25}$$

Similarly one discusses space representation of the integral of a product of field operators. Again ignoring ghost degrees of freedom we represent the integral of $A_1 * A_2 * \ldots * A_N$, defined as

$$\int A_1 * A_2 \cdots * A_N \equiv \int \prod_{i=1}^{N} \mathcal{D}x_i(\sigma) \prod_{i=1}^{N} \prod_{0\leq\sigma\leq\frac{\pi}{2}} \delta[x_i(\sigma) - x_{i-1}(\pi - \sigma)] A_i(x_i(\sigma))\ , \tag{26}$$

by means of a vector

$$\mid V_N\rangle \equiv V_N(a_1^\dagger, a_2^\dagger, \ldots, a_N^\dagger) \prod_{i=1}^{N} \mid 0\rangle_i \tag{27}$$

in the product Fock space of N strings. In this notation

$$\int A_1 * \cdots * A_N = \langle V_N \mid\mid A_1\rangle \cdots \mid A_N\rangle\ . \tag{28}$$

Again one represents $V_N(a_1^\dagger, a_2^\dagger, \ldots, a_N^\dagger)$ as an exponential form

$$V_N(a_1^\dagger, a_2^\dagger, \ldots, a_N^\dagger) = \exp\left[-\frac{1}{2}(a_i^\dagger \mid V_N^{ij} \mid a_j^\dagger)\right] \tag{29}$$

and derives the equations for the matrix $(V_N^{ij})_{nm}$. The most important are the three–point vertex V_3, which is the interaction term of Lagrangian, and V_4 which is useful for proving associativity. The cyclic symmetry of the overlap conditions allows for certain simplification. The three–string vertex is defined by the overlap equations

$$[x^r(\sigma) - x^{r-1}(\pi - \sigma)]\ |V_3>= 0\ , \qquad 0 \le \sigma \le \frac{\pi}{2}$$

$$[x^r(\sigma) - x^{r+1}(\pi - \sigma)]\ |V_3>= 0\ , \qquad \frac{\pi}{2} \le \sigma \le \pi \tag{30}$$

and the analogue equations for the momentum variables $p^r(\sigma), r = 1, 2, 3$. In terms of Z_3 linear combinations Q_a, the overlap conditions $x_r(\sigma) = x_{r-1}(\pi - \sigma)$ read

$$Q(\sigma) = \begin{cases} eQ(\pi - \sigma) & \text{for } 0 \le \sigma \le \frac{\pi}{2}\ , \\ \bar{e}Q(\pi - \sigma) & \text{for } \frac{\pi}{2} \le \sigma \le \pi\ , \end{cases} \tag{31}$$

and

$$Q_3(\sigma) = Q_3(\pi - \sigma)\ , \tag{32}$$

where we have denoted $Q(\sigma) \equiv Q_1(\sigma) = \overline{Q}_2(\sigma)$. Clearly one has a separation of variables and the vertex $\mid V_3\rangle$ takes the form

$$\mid V\rangle = \exp[-\frac{1}{2}(A_3^\dagger \mid C \mid A_3^\dagger) - (A^\dagger \mid U \mid \overline{A}^\dagger)] \mid 0\rangle_{1,2,3}\ , \tag{33}$$

where C_{nm} is the integration matrix and the matrix U satisfies the nontrivial overlap equations. This reduction from three to two essential degrees of freedom shows the high symmetry present in the midpoint overlap conditions. The equations are solved and the form for the vertex operator is constructed using the methods of conformal field theory [17]. This will be summarized shortly.

The Hilbert space language gives the following explicit form for the $*$ operation. Consider two states $\mid A\rangle$ and $\mid B\rangle$ which we place in Hilbert space 1 and 2. The state $\mid A * B\rangle$ living in Hilbert space 3 is given by

$$\mid A * B\rangle = {}_{12}\langle 0 \mid V_3(a_1\ , a_2\ , -Ca_3^\dagger) \mid A\rangle_1 \mid B\rangle_2 \mid 0_3\rangle\ . \tag{34}$$

Associativity of the $*$ operator is now easily shown (for Fock space states, and states generated by $*$ products of Fock space states).

One can, in general, construct the fermionic version of vertex operators in the theory. They follow from the overlap equations of the fermionic ghost coordinates

$$c_\pm(\sigma) = \sum_{n=-\infty}^{\infty} c_n e^{\pm in\sigma} = c(\sigma) \pm i\Pi_b(\sigma) \tag{35}$$

and

$$b_\pm(\sigma) = \sum_n b_n e^{\pm in\sigma} = \Pi_c(\sigma) \pm ib(\sigma)\ . \tag{36}$$

First for the integration we have the overlaps

$$[c(\sigma) + c(\pi - \sigma)] \mid I\rangle = 0\ ,$$

$$[b(\sigma) - b(\pi - \sigma)] \mid I\rangle = 0\ , \tag{37}$$

which are easily solved by

$$\mid I^o\rangle = e^{\sum_n b_{-n}(-)^n c_{-n}} \mid c_o = 0\rangle\ . \tag{38}$$

The ghost number of this state is $\frac{1}{2}$ and the required ghost number $(-3/2)$ is achieved by additional midpoint insertions. The unique form is

$$| I\rangle = b_+(\frac{\pi}{2}) b_-(\frac{\pi}{2}) \, | I^o\rangle \, . \tag{39}$$

Concerning the existence in the ghost sector of a fermionic formulation one might wonder whether they are totally equivalent for all the constructions involved. Such equivalence was in fact demonstrated. One can interchangeably use one or the other formulation, indeed, depending on the application one or the other may turn out to be more useful.

The construction of vertex operators associated with light-cone or Witten–type geometry is, in general, achieved using conformal field theory methods. Here the basic ingredients are the conformal mappings that map the upper half complex plane (or the circle) into the scattering geometry in question. These then allow the construction of Neumann functions giving the vertex. In what follows, we shall summarize the mappings relevant [17,19] for reaching Witten's geometry. They are generalizations of mappings introduced by Mandelstam for the light-cone theory; their nontriviality is due to a double sheeted structure and a presence of curvature.

Using the inverse mappings $z = z(\rho)$ it is then straightforward to construct the Neumann functions in question. In the bosonic case one only has a simple change of variables $N(\rho, \rho') = N(z(\rho), z(\rho'))$. In general, for the ghosts and the superstring theory, the construction is done following the rules of conformal field theory, a variable of conformal weight J transforms under conformal transformations as

$$O'_J(z') = \left(\frac{\partial z}{\partial z'}\right)^J O_J(z) \tag{40}$$

with the associated transformation for the correlation functions:

$$\langle O_J(\rho) \, O_{J'}\rangle = \left(\frac{\partial z}{\partial \rho}\right)^J \langle O_J(z) \, O_J(z')\rangle \left(\frac{\partial z'}{\partial \rho'}\right)^{J'} \, . \tag{41}$$

The bosonic string variables $X_\mu, c(\sigma), b(\sigma)$ have respective conformal weights $J = 0, -1, 2$ and these facts will determine their ρ-plane Neumann functions. The superstring theory with half-integer conformal weights introduces some additional interesting features. Here conformal field theory techniques are even more instrumental.

Consider first the two-string identification (relevant for V_2). There one takes $\alpha_1 = +1, \alpha_2 = -1$ and the locations of strings 1 and 2 are $z_1 = i, z_2 = -i$ and

$$\rho = \ln\left(\frac{z - i}{z + i}\right) - i\frac{\pi}{2} \, . \tag{42}$$

In practice the inverse mapping is directly relevant; parametrizing the two strings by

$$\rho = \begin{cases} \xi_1 + i\sigma_1 \\ -(\xi_2 + i\sigma_2) + i\pi \end{cases} \tag{43}$$

we have that

$$z|_r = z_r\left(\frac{1 + ie^{\varsigma_r}}{1 - ie^{\varsigma_r}}\right) \, . \tag{44}$$

On each string $\varsigma_r = \xi_r + i\sigma_r$ (the asymptotic states are reached for $\xi_r \to -\infty$). At $\xi_1 = \xi_2 = 0 : z|_1(\sigma) = z|_2(\pi - \sigma)$ which is the needed two–string identification. It is very simple to go from two strings to one and obtain the mapping relevant for the integration state $|I >$. One simply identifies $z \longleftrightarrow -z$ which means cutting the disk and gluing it into a cone with one string on it. This is totally equivalent to taking $z = \sqrt{\omega}$ and the mapping is then

$$\rho = \ln \frac{\sqrt{\omega} - i}{\sqrt{\omega} + i} - i\frac{\pi}{2} \, . \tag{45}$$

This corresponds to

$$\omega = -\left(\frac{1+ie^{\rho}}{1-ie^{\rho}}\right)^{\frac{1}{2}} \tag{46}$$

with a single–string parametrization on $\rho = \tau + i\sigma$. This clearly gives the required folding since for $\tau = 0$ one has

$$\omega(\sigma) = \omega(\pi - \sigma) \ . \tag{47}$$

The generalization of all this to higher numbers of strings is now obvious. In general for an even number of strings one uses the simple Schwarz-Christofell map and then one goes to the odd case by identification $z = -z$ or a simple additional map $z^2 = \omega$.

For Witten's three–string rearrangement, the mapping corresponds to that of six strings with alternating lengths ($\alpha_1 = 1,\ \alpha_3 = -1,\ \alpha_2 = 1,\ \alpha_{-1} = -1,\ \alpha_{-3} = +1,\ \alpha_{-2} = -1$) and it equals

$$\rho = \sum_{i=1}^{3}[\alpha_i \ \ln(z - z_i) + \alpha_{-i} \ \ln(z + z_i)] \ . \tag{48}$$

Imposing reflection symmetry ($z = -z$) on the six–string diagram leads to the three–string rearrangement process that we are after. The identification of reflected points means that we are actually mapping from the upper semicircle with points on the real line being sewed together. This means that for three strings we are mapping from a cone, and this is what it should be. Witten's geometry has a curvature at the midpoint and our dual model surface (the cone) has an analogue curvature. To summarize, therefore, the mapping is simply

$$\rho = \ln\left(\frac{z^3 - i}{z^3 + i}\right) - \frac{i\pi}{2} \ , \tag{49}$$

with the z-plane being a cone. For the construction of the vertex we shall need the Neumann function; it will simply be the Neumann function on the disk but with the identification $z = -z$ we have additional image charges at $-z'$ and $-1/\bar{z}'$. Consequently

$$N(z, z') = \ln|z^2 - z'^2||z^2 - \frac{1}{\bar{z}'^2}| \ . \tag{50}$$

As a final comment, one has that the half disk or the cone can be represented also by a further mapping $z = \sqrt{\omega}$. With this

$$\rho = \ln\left(\frac{\omega^{\frac{3}{2}} - i}{\omega^{\frac{3}{2}} + i}\right) - \frac{i\pi}{2} \tag{51}$$

and the mapping is now from the whole disk, but with a cut. In that case, one has the standard Neumann function on the whole disk. All of these representations are obviously equivalent. Some related forms were used by other authors [20-24].

More general conformal mapping are involved in four and general N-string Witten scattering processes. It is also easy to construct mappings that are intermediate between Witten and the light-cone geometries [25-28]. However, it does not mean that any vertex can be used for constructing a field theory. What one has to show is not only that the three string amplitudes come out correctly, but also the higher $N > 3$ point amplitudes. This, in general, leads to higher point vertices.

The amplitude for the scattering is in general given by the path integral

$$\int \mathcal{D}X_\mu(\rho) \exp\{-\frac{1}{4\pi}\int d^2\rho(\partial X)^2 + \sum_r \int_0^\pi d\eta_r X(\rho_r)P_r(\eta_r)\} \ , \tag{52}$$

where general external states are represented by the momentum distribution functions $P_r(\eta_r)$. The vertex is therefore essentially given by

$$\mathcal{V} = \exp\left\{-\frac{1}{2}\sum_{r,s}\int_0^{\pi} d\eta_r \int_0^{\pi} d\eta'_s P_r(\eta_r) N(\rho,\rho') P_s(\eta'_s)\right\} , \tag{53}$$

where the Neumann function is defined to obey

$$(\partial_\tau^2 + \partial_\sigma^2) N(\rho,\rho') = 2\pi\delta^{(2)}(\rho-\rho') .$$

After a Fourier decomposition and the separation of a diagonal wave function piece, we have

$$\mathcal{V} = \exp\left\{\frac{1}{2}\sum_{r,s=1}\sum_{n,m=0}^{\infty} P_n^r N_{nm}^{rs} P_m^s\right\} . \tag{54}$$

The Fourier coefficients are defined by

$$N_{nm}^{rs} - \delta_{rs}\frac{\delta_{nm}}{n} e^{-n|\xi_r-\xi'_s|} = \frac{2}{\pi^2} e^{-m\xi_r - n\xi'_s}\int_0^{\pi} d\eta_r \int_0^{\pi} d\eta'_s \cos m\eta_r \cos n\eta'_s N(\rho,\rho'). \tag{55}$$

The Neumann function for the scattering process in question $N(\rho,\rho')$ is in general nontrivial but it is related to the simpler Neumann function of the upper half plane by a conformal transformation. The defining equation after a change of integration variables gives

$$N_{mn}^{rs} = \frac{1}{(2\pi)^2}\oint_{zr} dz \oint_{zs} dz' \frac{1}{(z-z')^2}\exp\{-m\varsigma_r(z) - n\varsigma_s(z')\} . \tag{56}$$

The Fourier coefficients N_{nm}^{rs} were obtained in a closed form using some related techniques. The interaction vertex takes the form

$$| V_3\rangle = \exp\left[\frac{1}{2}\sum_{r,s}\sum_{n,m}\alpha_{-n}^r N_{nm}^{rs}\alpha_{-m}^s\right] | 0\rangle_{123} \tag{57}$$

and the three–string Neumann coefficients N_{nm}^{rs} are given in terms of basic six–string coefficients $N_{nm}^{r,s}$ as

$$N_{nm}^{rs} = N_{nm}^{r,s} + N_{nm}^{r,-s} . \tag{58}$$

For the ghost interaction vertex one has

$$| V_3^{gh}\rangle = \exp\left[\sum_{r,s}\sum_{n,m} b_{-n}^r \tilde{N}_{nm}^{rs} m c_{-m}^s\right] \tag{59}$$

with $\tilde{N}_{nm}^{rs} = N_{nm}^{r,s} - N_{nm}^{r,-s}$. It corresponds to the overlap equations

$$\begin{aligned} &\left(c^r(\sigma) + c^{r-1}(\pi-\sigma)\right) = 0 \\ &\left(b^r(\sigma) - b^{r-1}(\pi-\sigma)\right) = 0 \ . \end{aligned} \tag{60}$$

The general structure of the Neumann coefficients agrees with the general structure following from the δ-function overlap equations. One of the main features that we have established for the Neumann function expression of the vertex is that it solves and explicitly represents the δ-

function interaction. This can be seen from the form and transformation properties of the basic functions $z/r = z(\rho)$ that appear in the Neumann function. More directly, we have verified that the Neumann coefficients and the vertex matrices U_{nm} obey the matrix overlap equations. This verification, on one hand, assures that there are no superficial midpoint insertions and, on the other, establishes a transition between the second (δ-function interaction) and first (path integral) quantization.

Concerning the overlap equations, one should note that rather nontrivial phenomena can occur at the midpoint. We have already mentioned the midpoint K_n anomalies. The overlap equations also develop anomalies of the midpoint $\sigma = \pi/2$. For example, naively one would expect the $p^r \frac{\pi}{2} = 0$ on the three-string vertex, however, these diverge as we approach the midpoint. In general, for an operator of conformal weight J one has the following behavior:

$$O_J(\frac{\pi}{2} - \epsilon) = \left(\frac{\partial\omega}{\partial\rho}\right)^J . \tag{61}$$

For the integration

$$\frac{\partial\omega}{\partial\rho} = \frac{z}{i}(\omega + 1) \sim \epsilon , \tag{62}$$

and for the interaction vertex

$$\frac{\partial\omega}{\partial\rho} = \frac{i}{3}z^{-1}(\omega^3 + 1) \sim \epsilon^{-\frac{1}{3}} , \tag{63}$$

and these induce the above-mentioned anomalies.

The conformal field theory techniques can be used to construct vertex operators on generalized vacuum states [18] (this will be especially useful for superstrings). Here one can consider the integration, where particularly relevant is the construction based on the $SL(2,R)$ state $|\Omega>$. It is defined by

$$b_n \ |\Omega>= 0 , \qquad n \geq -1 ,$$

$$c_n \ |\Omega>= 0 , \qquad n \geq 2 , \tag{64}$$

and has just the ghost number $N_{gh} = -\frac{3}{2}$ required for the needed for the $|I>$ state. On this vacuum we have the form

$$|I^{cb}>= \exp\{ \sum_{\substack{n\geq 2\\ m\geq -1}} b_n\tilde{C}_{nm}c_{-m}\}|\Omega\rangle . \tag{65}$$

with <u>no</u> additional midpoint insertions. The Neumann function is taken so as to obey the correct overlaps and have the correct midpoint behavior at $\sigma = \pi/2$. It is

$$\tilde{C}(\rho,\rho') = \left(\frac{\partial\omega'}{\partial\rho}\right)^{-1} \frac{1}{\omega - \omega'} \left(\frac{\partial\omega'}{\partial\rho'}\right)^2 \cdot \left(\frac{\omega+1}{\omega'+1}\right)^3 . \tag{66}$$

Apart from the factor $\left(\frac{\omega+1}{\omega'+1}\right)^3$, this is the standard $< c\, b >$ correlation function and the additional factor is simply there to assure the appropriate Fourier expansion required for our vacuum state $|\Omega>$.

The quadratic form appearing in the three–ghost vertex can also be given by conformal field theory with the following $< cb >$ correlation function:

$$N^{cb}(\rho,\rho') = \left(\frac{\partial\omega}{\partial\rho}\right)^{-1} \frac{1}{\omega - \omega'} \left(\frac{\partial\omega'}{\partial\rho'}\right)^2 \frac{\omega^3 + 1}{\omega'^3 + 1} . \tag{67}$$

The additional factor appearing is inserted to give the Fourier expansion:

$$N^{cb}(\rho,\rho') = \sum_{\substack{n\geq 0\\ m\geq 1}} e^{in\sigma}\, e^{im\sigma'} \tilde{N}^{ab}_{nm} m + \delta_{a,b} \sum_{n\geq 0} e^{in(\sigma-\sigma')} \ , \tag{68}$$

which is appropriate for our $|+\frac{1}{2}>$ vacuum state.

The fermionic formulation of ghosts was used to prove directly the BRST invariance of the theory. For the three–string vertex, one first has the K_n-invariance

$$(K_n^1 + K_n^2 + K_n^3)\,|V_3\rangle = 0 \ . \tag{69}$$

Here the result follows after cancellation of bosonic and fermionic midpoint anomalies which are found to equal

$$K_{2N}^{x} = \frac{D\cdot 5}{2\cdot 3^2}\, N(-)^N$$

$$K_{2N}^{cb} = -\frac{13\cdot 5}{3^2}\, N(-)^N \tag{70}$$

and clearly cancel when $D = 26$.

The proof of BRST invariance is straightforward and goes as follows. One writes the total BRST charge

$$Q = Q^1 + Q^2 + Q^3 = \sum_{r=1}^{3}\Big(\sum_{n=1}^{\infty}\Big[c_{-n}^{n}\mathcal{L}_n^r + \mathcal{L}_{-n}^r c_n^r\Big] + c_0^r(\mathcal{L}_0^r - 1)\Big) \ , \tag{71}$$

where $\mathcal{L}_n = L_n^x + \frac{1}{2}L_n^{gh}$. From the vertex there follow the overlap equations

$$(c_n^r - \tilde{N}_{nm}^{rs}\, m\, c_{-m}^s) = 0$$

$$(b_n^r + n\tilde{N}_{nm}^{rs}\, b_{-m}^s) = 0 \tag{72}$$

and similarly for the L_n operators

$$(L_n^r + n\tilde{N}_{nm}^{rs}\, L_{-m}^s - k_n) = 0 \ . \tag{73}$$

They, like b, have conformal weight $J = 2$ and obey identical equations (apart from the k_n anomaly). Using the above set of equations, one easily finds

$$Q\,|V_3^x\rangle\,|V_3^{cb}\rangle = \sum_{r=1}^{3}\sum_{n} c_{-n}^r(k_n^x + k_n^{gh}) = 0 \ . \tag{74}$$

Consequently the BRST symmetry is seen to follow from the cancellation of k_n-anomalies and the overlap equations. This is a general phenomenon.

On-shell amplitudes of the new vertex show complete agreement with dual models. This agreement follows from the on-shell conformal invariance of the above construction. Namely the fact that we have related the Witten and the unit disk geometries through a conformal map implies that there exists an operator conformal transformation.

$$|V_3^{witten}> = e^{\sum_n a_n L_n}\,|V_3^{CSV}> \tag{75}$$

to the dual model CSV vertex operator. This transformation was constructed with a systematic recursion formula for the coefficients (a_n) which are basically related to the mapping function $\rho = \rho(z)$.

In conclusion, in the present approach all the basic ingredients of the theory are expressed in the string Hilbert space. The equation of motion for the string field is given by

$$Q|A\rangle + {}_{12}\langle 0|V_3(a_1\ , a_2\ , -Ca_3^+)\ |A\rangle_1\ |A\rangle_2\ |0\rangle_3 = 0\ , \tag{76}$$

and one can contemplate finding "classical" solutions by appropriate ansatzes in the Hilbert space.

III. Superstring Field Theory

The outline of the open superstring field theory goes as follows. In the Neveu-Schwarz-Ramond framework one has, in addition to the coordinates $x_\mu(z)$, also the fermionic variable $\psi_\mu(z)$ and, in addition to the ghost variables c, b the bosonic superghost variables $\gamma(z), \beta(z)$. In the Neveu-Schwarz sector the variables are half-integer moded, while for Ramond, one has integer moding; the superghost zero modes γ_o, β_o in general play an important role. The physical states of the Neveu-Schwarz sector $| A_{NS} >$ have $N_{gh} = -\frac{1}{2}$ coming from the ghosts while in the Ramond sector $| A_R >$ is taken to have ghost number $N_{gh} = 0$ (here the $-\frac{1}{2}$ ghost number of the ghosts is cancelled by a $+\frac{1}{2}$ contribution assigned to superghosts). The string field $A = (A_{NS}, A_R)$ then has ghost number $\left(-\frac{1}{2}, 0\right)$ and the free action reads

$$S_o = < A_{NS} \mid Q \mid A_{NS} > + < A_R \mid Y\,(\pi/2)\, Q \mid A_R > , \tag{77}$$

with Y being the inverse picture–changing operator.

Witten has generalized his $\int$ and $*$ operations to superstrings in the following way. For the integration one defines

$$\oint (A_{NS}, A_R) = \int Y\,(\pi/2)\, A_{NS}\ , \tag{78}$$

where on the right hand side one still has a folding of a string about $\pi/2$. The generalized star product reads

$$\begin{aligned}(A_{NS}, A_R) \star (A'_{NS}, A'_R) &= X\,(\pi/2)\,(A_{NS} * A'_{NS}) + A_R * A'_R \\ &\quad X\,(\pi/2)\,(A_{NS} * A'_R + A_R * A'_{NS})\ .\end{aligned} \tag{79}$$

Here X is the picture–changing operator. In terms of the generalized $\star$ and $\oint$ operations the gauge invariant action for open superstrings takes the same form as for the bosonic string

$$S = \oint \left(A \star A + \frac{2}{3} A \star A \star A\right). \tag{80}$$

This Lagrangian is again gauge invariant under $\delta A = Q\Lambda + A \star \Lambda - \Lambda \star A$, where the gauge parameter $\Lambda = (\Lambda_{NS}, \Lambda_R)$ carries ghost number $\left(-\frac{3}{2}, -1\right)$. The theory is also invariant under supersymmetry transformations

$$\begin{aligned}\delta A_{NS} &= Q_\alpha A_R \\ \delta A_R &= X\left(\frac{\pi}{2}\right) Q_\alpha A_{NS}.\end{aligned} \tag{81}$$

The supersymmetry charges are to be given by the fermion emission vertex

$$Q_\alpha = \int do S_\alpha\,(\sigma)\, e^{-\frac{\phi}{2}}\ , \tag{82}$$

where $S_\alpha(\sigma)$ is the spin field and ϕ is the bosonized superghost.

The operator construction which follows represents the Lagrangian as

$$
\begin{aligned}
S = &< A_{NS} \mid Q \mid A_{NS} > + < A_R \mid YQ \mid A_R > \\
&+ \frac{2}{3} < V_3^{NSNSNS} \parallel A_{NS} >\mid A_{NS} \mid A_{NS} > \\
&+ 2 < V_3^{NSRR} \parallel A_{NS} >\mid A_R >\mid A_R >
\end{aligned}
\tag{83}
$$

with the explicit forms for the $NS - NS - NS$ and also the $NS - R - R$ vertex operators. In this representation on–shell three–point amplitudes are directly computable and the BRST symmetry can be directly verified.

One also has the two-vertex operators corresponding to $NS - NS, R - R$ and the $NS - R$ string overlaps. The Neveu-Schwarz and the Ramond two-vertex states $|V_2^{NSNS} >$ and $|V_2^{RR}\rangle$ represent scalar products and define the hermiticity properties of the corresponding string fields, which read

$$
\begin{aligned}
< A_{NS}| &=< V_2^{NSNS}|A_{NS} > \\
< A_R| &=< V_{2\ o}^{RR}|A_R > .
\end{aligned}
\tag{84}
$$

In terms of two-string overlaps the free action can be written as

$$
S_o =< V_2^{NSNS}||A_{NS} > |QA_{NS} > - < V_2^{RR}|A_R > |QA_R > \tag{85}
$$

with $|V_2^{RR} >= Y(\pi/2)|V_{20}^{RR} >$. These scalar product operators have a simple form. The mixed $NS - R$ two-string overlap vertex $|V_2^{NSR}\rangle$ also exists and is nontrivial. It is relevant for giving an alternative form for supersymmetry transformations.

Finally, we have the integration state $|I >$. In superstring field theory one only defines an integration for Neveu-Schwarz strings $|I_{NS} >$:, there is no need for a Ramond integration due to the form of Witten's generalized integration $\oint$ which has no Ramond contribution.

The construction reported here summarizes Ref. [18]. Some similar results were derived in [23]. There is, furthermore, the work of [21] which is different; relying on bosonization of all fermionic degrees of freedom the superstring problem can be transcribed into a bosonic one.

The procedure followed in general to obtain the vertices was to perform a conformal transformation that mapped the diagram to the upper half-plane. On the upper half-plane the scattering amplitude of free strings coupled by the vertex could easily be constructed using the methods of conformal field theory. Compared with the bosonic string where the variables x_μ, c, b had integer conformal weights $J = 0, -1$ and 2, in the present case we have half-integer conformal weights $J_\psi = \frac{1}{2}$, $J\gamma = -\frac{1}{2}, J_\beta = \frac{3}{2}$. A conformal field transforms in general as

$$
\psi'_J(z') = \left(\frac{\partial z}{\partial z'}\right)^J \psi_J(z) \ . \tag{86}
$$

The overlap condition for a general string variable can then be deduced by equating its value at z with its value at $z' = -\frac{1}{z}$, (this being the conformal transformation that corresponds to $\sigma \to \pi - \sigma$ with $\rho = \ln z$). Thus the overlap condition for the identity for a variable of conformal weight J reads

$$
\psi_J(\sigma) = (-)^J \psi_J(\pi - \sigma). \tag{87}
$$

For the Neveu-Schwarz variables then

$$\begin{aligned}(\psi_+(\sigma) - i\psi_+(\pi-\sigma)) \mid I\rangle = 0, \qquad (0 \le \sigma \le \frac{\pi}{2}),\\ (\psi_-(\sigma) + i\psi_-(\pi-\sigma)) \mid I\rangle = 0, \qquad (0 \le \sigma \le \frac{\pi}{2}).\end{aligned} \tag{88}$$

The opposite signs in the ψ_+ and ψ_- equations are dictated by the boundary conditions.

In the case of the bosonic string the overlap conditions implied that the Virasoro generators obeyed $L_\pm(\sigma) = L_\pm(\pi-\sigma)$. The Fourier components of this equation, $K_n \equiv L_n - (-1)^n L_{-n}$, were derivations of the integration defined by $|I\rangle$. The supersymmetric string has a supersymmetric generalization of the Virasoro algebra, which contains the fermionic charges G_r. The fermionic charge density is given by $G_\pm(\sigma) = P_\pm(\sigma)\cdot\psi_\pm(\sigma)$ and the overlap conditions for ψ and P imply that

$$G_\pm(\sigma) \mp iG_\pm(\pi-\sigma) = 0 \tag{89}$$

in addition to the $L(\sigma)$ equation. Correspondingly, we have a subalgebra of the full superconformal algebra that annihilates $|I\rangle$.

The Neveu-Schwarz part of the integration state $|I>$ is expressed as

$$|I^\psi\rangle = \exp[\frac{1}{2}\sum_{s,r\ge\frac{1}{2}} \psi_{-r} I_{rs} \psi_{-s}]|0> \tag{90}$$

with a standard Fock vacuum $|0>$. The overlap equations for the antisymmetric matrix I_{rs} can again easily be worked out; the solution is obtained in terms of the corresponding Neumann function.

On the circle the Neumann function is

$$K(\omega,\omega') = \langle\psi(\omega)\psi(\omega')\rangle = \frac{1}{(\omega-\omega')}. \tag{91}$$

We transform back to the strip using the transformation properties of Ψ and

$$K(\rho,\rho') = \left(\frac{\partial\omega}{\partial\rho}\right)^{\frac{1}{2}} \frac{1}{\omega-\omega'} \left(\frac{\partial\omega'}{\partial\rho'}\right)^{\frac{1}{2}}, \tag{92}$$

Explicitly using the mapping relevant to the integration with $\omega = z^2$ we have

$$K(\rho,\rho') = \left(\frac{z^2+1}{i}\right)^{\frac{1}{2}} \frac{\sqrt{z}\sqrt{z'}}{z^2-z'^2} \left(\frac{z^2+1}{i}\right)^{\frac{1}{2}}. \tag{93}$$

The expansion of this Neumann function in a Fourier series reads

$$K(\rho,\rho') = \sum_{r=\frac{1}{2}}^{\infty} e^{-ir(\sigma-\sigma')} + \sum_{r,s=\frac{1}{2}}^{\infty} e^{ir\sigma} e^{is\sigma'} I_{rs}. \tag{94}$$

The first term corresponds to the propagation of free strings, while the second scattering coefficients of plane waves, $e^{ir\sigma}e^{is\sigma}$, give the integration vertex coefficents I_{rs}. The overlap equations follow from the fact that the Neumann function itself satisfies the appropriate overlaps, namely when $\rho = i\sigma$

$$K(\sigma,\sigma') = \begin{cases} iK(\pi-\sigma,\sigma'), & o \le \sigma \le \pi/2\,, \\ -iK(\pi-\sigma,\sigma'), & \pi/2 \le \sigma \le \pi\,. \end{cases} \tag{95}$$

On the integration state one has the representation

$$\psi_+(\sigma)|I^\psi\rangle = \int \frac{d\sigma'}{2\pi} K(\sigma,\sigma')\psi_\mp^{cr}(\sigma')|I^\psi\rangle\,. \tag{96}$$

The operator overlap equations then follow from the matching conditions obeyed by the Neumann function $K(\sigma,\sigma')$.

Let us now discuss the scalar product operation $|V_2>$. For example, for two Neveu-Schwarz strings one has the scalar product

$$\int A^1 * A^2 =< A^1| < A^2|V_2^{NSNS}> \tag{97}$$

with the NS-NS vertex obeying $[\psi^1_\pm(\sigma) \pm i\psi^2_\pm(\pi-\sigma)]|V^2_{NS,NS}\rangle = 0$. In this case there is no singularity at the midpoint, so the above overlap equation holds for all $0 \leq \sigma \leq \pi$. The solution is simple, namely

$$|V^2_{NS,NS}\rangle = \exp\left[i\sum_{r\geq\frac{1}{2}} \psi^{1\mu}_{-r}(-)^r\psi^{2\mu}_{-r}\right]|0\rangle_{12}\,. \tag{98}$$

This quadratic form also trivially follows from the Neumann function

$$K^{ab}(\sigma,\sigma') = \left(\frac{\partial z}{\partial\varsigma_a}\right)^{\frac{1}{2}}\frac{1}{z-z'}\left(\frac{\partial z'}{\partial\varsigma'_b}\right)^{\frac{1}{2}} \tag{99}$$

with the two-string mapping $\rho = ln\frac{z-i}{z+i} - i\frac{\pi}{2}$ and the parametrization $\rho = \varsigma_1$ on string 1 and $\rho = -\varsigma_2 + i\pi$ on string two; for each string $\varsigma = \tau + i\sigma$. An identical form follows for the R-R scalar product and also the superghost degrees of freedom.

The three-string vertex is written in the form

$$|V_3^\psi>= \exp\left\{\frac{1}{2}\sum_{r,s\geq 1/2}\psi_r^{\mu(a)}K^{ab}_{rs}\psi_s^{\mu(b)}\right\}|0>_{123}\,, \tag{100}$$

and it corresponds to the overlap equations

$$\psi^a_\pm(\sigma) = \pm i\psi^{a-1}(\pi-\sigma) \qquad a = 1,2,3\ldots\,. \tag{101}$$

The quadratic form is given by

$$\triangle^\psi = \int\frac{d\sigma}{2\pi}\int\frac{d\sigma'}{2\pi}\psi^{(a)}_{an}(\sigma)K^{ab}(\sigma,\sigma')\psi^{(b)}_{an}(\sigma')\,, \tag{102}$$

where $K^{ab}(\sigma,\sigma')$ represents the correlation function of a fermion on the ath and a fermion on the bth string. The Fourier coefficients K^{ab}_{rs} are obtained from the explicit form of $K(\rho,\rho\prime)$ using the three–string mapping $\rho = \ln\frac{\omega^{\frac{3}{2}}-i}{\omega^{\frac{3}{2}}+i} - i\frac{\pi}{2}$. The basic function entering in these expansions is $g(\sigma) = \omega_a^{1/4}\left(\frac{1+ie^{i\sigma}}{1-ie^{i\sigma}}\right)^{1/6}$. This is recognized as the mapping appropriate for a 12-string rearrangement and we have the twelve string analogues of our earlier six–string and three–string coefficients a_n, b_n. The coefficients K^{ab}_{rs} are explicitly given in Ref. [18]. From eq.(26) we have the integral formula

$$K^{ab}_{rs} = \int_{\omega_a}\frac{d\omega}{2\pi i}\int_{\omega_b}\frac{d\omega'}{2\pi i}e^{-n\rho}e^{-m\rho'}\left(\frac{d\omega}{d\rho}\right)^{-1/2}\left(\frac{d\omega'}{d\rho'}\right)^{-1/2}\frac{1}{\omega-\omega'} \tag{103}$$

which can be used for various general considerations.

For the Ramond sector one has a similar construction. Of main interest is the interaction operator for $R-R-NS$ strings. This and the NS identity are sufficient to give all the Ramond terms in the action. Namely, the Ramond kinetic term is constructed through $\langle I_{NS}|V^3_{RRNS}\rangle = |V^2_{RR}\rangle$, and the vertex operator itself gives the coupling of Ramond and Neveu-Schwarz fields.

It is given by the following overlap equations:

$$\begin{aligned}
&\left(\chi^1_\mp(\sigma) \mp i\psi^3_+(\pi-\sigma)\right)|V^3_{RRNS}\rangle = 0 \\
&\left(\chi^2_\pm(\sigma) - i\chi^1_\pm(\pi-\sigma)\right)|V^3_{RRNS}\rangle = 0 \qquad (0 \le \sigma \le \frac{\pi}{2}) \\
&\left(\psi^3_\pm(\sigma) - i\chi^2_\pm(\pi-\sigma)\right)|V^3_{RRNS}\rangle = 0.
\end{aligned} \tag{104}$$

The signs are consistent with the boundary conditions on Ramond strings χ^1, χ^2 and the Neveu-Schwarz string ψ^3. The quadratic form for the vertex is again given by the Neumann function, which is now constructed as

$$R(\rho,\rho') = \frac{1}{2}\left(\frac{\partial\omega}{\partial\rho}\right)^{1/2}\frac{1}{\omega-\omega'}\left(\frac{\partial\omega'}{\partial\rho'}\right)^{1/2}\left[\left(\frac{(\omega-\omega_1)(\omega'-\omega_2)}{(\omega'-\omega_1)(\omega-\omega_2)}\right)^{1/2} + \left(\frac{(\omega'-\omega_1)(\omega-\omega_2)}{(\omega-\omega_1)(\omega'-\omega_2)}\right)^{1/2}\right]. \tag{105}$$

The additional factors are there to insure the integer–moded Fourier expansions appropriate for Ramond strings.

All the foregoing constructions apply also to superghosts. First one has the bosonized form where the commuting superconformal ghost $\gamma(\sigma), \beta(\sigma)$ operators are replaced by two scalar fields $\phi(\sigma)$ and $\overline{\phi}(\sigma)$. In the bosonized approach no additional constructions are needed. The vertex operators for the new degrees of freedom are the same as the bosonic coordinate operators. The only additional effects that have to be taken into account are the insertions at the midpoint. This bosonized approach is quite elegant since it explicitly exhibits the conformal and superconformal anomalies of the Riemann surfaces in question and was given in detail in [18]. We also have a formulation of the theory in terms of the basic ghost coordinates b, c and β, γ.

It is well known that the (super)ghost system γ, β allows an infinite set of different vacuum states. These "picture changed" vacua labelled by $|0;n\rangle$ are characterized by different ghost numbers $N_g = n$ and different sets of annihilation (creation) operators. We gave a formulation of the theory in this more general representation. This formulation is useful for various applications, one being manifest BRST symmetry. In general it is in the Neveu-Schwarz sector that one needs the picture changing. In the Ramond sector things are actually simpler, for instance, there is no Ramond integration and no picture changing insertion $X(\frac{\pi}{2})$ in the NS-R-R vertex. Of special interest is the construction based on the $SL(2,R)$ invariant vacuum state $|\Omega\rangle = |0;1\rangle$. Its creation operators are given by

$$\left.\begin{matrix}\beta_{-r} & r \ge \frac{3}{2} \\ \gamma_{-s} & s \ge -\frac{1}{2}\end{matrix}\right\} \quad \Omega \tag{106}$$

and it gives the simplest form for the superghost integration state $|\,I^{\gamma\beta}\rangle$. This is achieved by the correlation function

$$K^{(+1)}(\rho,\rho') = \left(\frac{\partial\omega}{\partial\rho}\right)^{-1/2}\frac{1}{\omega-\omega'}\left(\frac{\partial\omega'}{\partial\rho'}\right)^{3/2}\left(\frac{\omega+1}{\omega'+1}\right)^2, \tag{107}$$

which corresponds to the canonical $<\gamma\beta>$ correlation function with an additional factor. The complete superghost integration state was given then by

$$|\,I^{\gamma\beta}\rangle = Y(\frac{\pi}{2})\exp\Big(\sum_{\substack{r\ge 3/2\\ s\ge -1/2}}\beta_{-r}K^{(+)}_{rs}\gamma_{-s}\Big)\,|\,\Omega\rangle\,. \tag{108}$$

There are several other representations for the superghost integration state that we have constructed. One can perform a construction choosing any of the $| 0; n\rangle$ states as the vacuum; in general the quadratic form will be different but the insertions at the midpoint will also change. Of particular additional interest are the constructions using the states $| 0; 0\rangle$ and the $SL(2, R)$ conjugate state $| \overline{\Omega}\rangle = | 0; -1\rangle$. Either of the forms for the integration state is appropriate for demonstrating the BRST invariance. The discussion is simplest in the $SL(2, R)$ formulation. Similar considerations hold for the superghost interaction vertex.

We now summarize the construction of supersymmetry generators in Witten's theory using NS-R overlaps [29]. These give a natural form for the generators in which the equations necessary for supersymmetry are established. As such this offers an alternative to the bosonized spin field formulas. In addition, one has a precise derivation of the fermion emission vertex from Neveu-Schwarz and Ramond string overlaps.

One has a one parameter set of overlap equations which correspond to the parameter of the emission vertex. Consider the fermionic variables defined on the whole range $-\pi \leq \sigma \leq \pi$ by

$$\psi(\sigma) = \begin{cases} \psi_-(\sigma) & 0 \leq \sigma \leq \pi \\ \psi_+(-\sigma) & -\pi \leq \sigma \leq 0 \end{cases} \tag{109}$$

and similarly for χ. We define the general NS-R overlap $| V_2^{NSR}(\theta)\rangle$ to obey the following equations:

$$\begin{aligned} \psi^1(\sigma) + i\chi^2(\pi - \sigma) &= 0, \quad -\pi \leq \sigma \leq \theta\ , \\ \psi^1(\sigma) - i\chi^2(\pi - \sigma) &= 0, \quad \theta \leq \sigma \leq \pi\ . \end{aligned} \tag{110}$$

Again due to a change of sign at $\sigma = \theta$ one has assured that the NS and R boundary conditions are obeyed, the above set of equations is consistent and has a unique solution in terms of a quadratic form. After a trivial twisting transformation these are seen to be related to the coefficients defining the fermion emission vertex. The overlap equations are especially useful for discussing supersymmetry in the operator form of Witten's superstring field theory. One can write the transformations in the form

$$\begin{aligned} \delta \mid A_{NS}\rangle_1 &= \bar{\epsilon}\ \hat{S}_{12}\rangle_1 \mid A^R\rangle_2 \\ \delta \mid A_R\rangle_1 &= \bar{\epsilon}\ \hat{S}_{12}^+ \mid A^{NS}\rangle_2 \end{aligned} \tag{111}$$

with the supersymmetry generators given by

$$\hat{S} \equiv {}_2\langle S_{RNS}\rangle_1 = {}_{32}\langle V_2^{NSR} \mid V_2^{NSNS}\rangle_{31} \tag{112}$$

and the conjugate

$$\hat{S}^+ = {}_2\langle S_{NSR}\rangle_1 = {}_{32}\langle V_2^{RNS} \mid X \mid V_2^{RR}\rangle_{31}\ . \tag{113}$$

The invariance of the action under the supersymmetry transformations defined above goes as follows. First for the free part of the action one finds that the variation equals

$$\delta S_0 = 2\bar{\epsilon}\langle V_2^{NSR} \mid (Q_{NS}^1 + Q_R^2) \mid A^{NS}\rangle \mid A^R\rangle\ , \tag{114}$$

which is zero because the two–string NS-R overlap vertex $| V_2^{NSR}\rangle$ is constructed to be BRST invariant. Actually, a slightly more precise definition of supersymmetry transformations is needed which takes into account the fact that in the Ramond sector some operators like Y have zero modes. In general, one has to project these out. A complete understanding of these issues is given in [30]. For the invariance of the interaction term one obtains necessary relationships between the NS-NS-NS and R-R-NS vertex operators. Namely the condition is

$$ {}_{423}\langle V_3^{NSNSNS} \mid \hat{S}_{14} + {}_{143}\langle V^{RRNS} \mid \hat{S}^+_{24} = 0 \,. $$

Here in the first NS-NS-NS vertex the $\hat{S}$ operator converts the configuration into a R-NS-NS overlap while in the second R-R-NS term one changes the second Ramond string (into NS) obtaining again a R-NS-NS overlap. Consequently by adjusting a sign one has the required cancellation. A more detailed analysis is given in [29].

In a recent paper [44] it was also pointed out that one needs some singular four–point contact terms in the superstring interaction. These are associated with the fact that the operator insertions of X in the amplitudes receive contributions from the singular operator product. Contact four–point terms seem to be generally present in the superstring interaction [45].

In summary, we have that the open string field theory is completely formulated and, as described above, explicitly constructed. It gives a gauge invariant description of open string scattering amplitudes.

IV. Closed Strings, Open Questions

We review now some further developments of string field theory. The first subject which is of direct relevance is the issue of gauge fixing in Witten's theory (and the gauge invariant string field theory in general). This problem is nontrivial with some basic open questions.

In practice, one has a simple guess for the gauge fixed theory: it is represented by the same Lagrangian but without the ghost number restriction on the string field [26]. Gauge invariance is then replaced by the (second quantized) BRST invariance and one has a clear analogy with the Polyakov string type Feynman rules.

Consider first the gauge fixing procedure for the free theory. The linearized gauge invariance $\delta A = Q\Lambda$ (where Λ has ghost number $-\frac{3}{2}$) is fixed by imposing the gauge condition $b_0 A = 0$ by introducing the Faddeev-Popov ghost and antighost fields $\Phi^{(-3/2)}, \Phi^{(1/2)}$ and the Lagrangian is

$$ S_0^{(1)} = \int \left(A * QA + \Phi^{(1/2)} * Q\Phi^{(-3/2)} \right) . \tag{115} $$

Clearly one now has a further invariance $\delta\Phi^{(-3/2)} = Q\Lambda^{(-5/2)}$ and after fixing $b_0\Phi^{(-3/2)} = 0$ one ends up with additional ghosts and antighost fields $\Phi^{(-5/2)}, \Phi^{(3/2)}$. This continues and the final gauge fixed Lagrangian is then

$$ S_{0_{GF}} = \int A * QA + 2\sum_{n=0}^{\infty} \int \Phi_{n+1/2} * Q\Phi_{-n-3/2} \,. \tag{116} $$

Denoting $A = \Phi^{(-1/2)}$ and after combining all the fields into

$$ \hat{A} = \sum_{n=-\infty}^{\infty} \Phi^{(n-1/2)} \tag{117} $$

one has the same Lagrangian as in the gauge invariant case

$$ S_{0_{GF}} = \int \hat{A} * Q\hat{A} - \int (b_0\lambda) * \hat{A} \tag{118} $$

but with a gauge condition $b_0\Phi = 0$ enforced by a Lagrange multiplier and with no restriction on the gauge field $N(\hat{A}) \neq -\frac{1}{2}$ (all ghost number fields are present).

The above phenomenon (ghosts of ghosts) obviously comes from the fact that the gauge symmetry of the free theory is not simply given by $\delta A = Q\Lambda$ with arbitrary Λ. Because of nilpotency of Q those $\Lambda's$ which are of the form $\Lambda^{(1)} = Q\Lambda^{(2)}$ are ineffective and should not be counted. In this one should next ignore those $\Lambda^{(2)}$'s that are themselves equal to $Q\Lambda^{(3)}$ and so on to infinity. Consequently the parameter space of the free gauge group is slightly more involved but with the above specification of the zero modes ($\Lambda^{(1)}, \Lambda^{(2)}$, *etc.*) the precise nature of the gauge group is clear in free string field theory.

The situation is not so clear in the interacting case. It is not obvious that the nonlinear gauge transformations $\delta A = Q\Lambda + A * \Lambda - \Lambda * A$ in general allow zero modes. However, if one uses the equations of motion,

$$QA + A * A = (Q + A) * (Q + A) = 0 \tag{119}$$

and the above phenomena of ghosts might persist on shell.

A gauge fixing procedure which has worked for some on–shell–type gauge symmetries is due to Batalin and Vilkovisky and was applied to the string problem systematically in Ref. [31]. Based on this, the complete BRST invariant gauge fixed action [26,32] can be derived. The final Lagrangian turns out to be again of Witten form

$$S_{GF} = \int (\hat{A} * Q\hat{A} + \frac{2}{3}\hat{A} * \hat{A} * \hat{A} + b_0\lambda * \hat{A}) \tag{120}$$

with the fact that $\hat{A}$ has components with any ghost number (A gives $-\frac{1}{2}$, the ghosts give $N_g \leq -\frac{3}{2}$ and the antighosts $N_g \geq \frac{1}{2}$). The significant feature of the above gauge fixed action is the existence of second quantized BRST symmetry (the requirement for such symmetry is the main guiding principle involved in the $B - V$ scheme). Here the field transformations are

$$\begin{aligned} S\hat{A}_- &= (Q\hat{A} + \hat{A}^2)_- \\ S\hat{A}_+ &= -(b_0\hat{A})_+ \\ S\lambda &= 0. \end{aligned} \tag{121}$$

In addition to the invariance of the Lagrangian it was also established that the measure (in the functional integral) is also invariant [33]. This is an on–shell BRST invariance since one can show that

$$S^2 = ([\hat{A}, Q\hat{A} + \hat{A}^2])_- = 0 \ , \tag{122}$$

which holds only if the equations of motion $Q\hat{A} + \hat{A}$ are imposed. This is not surprising since the whole procedure for reaching the above gauge fixed action involved using the equations of motion. Related to the gauge fixing, one might wonder about the canonical and constraint structure of the gauge invariant Lagrangian. Some progress was made on this nontrivial problem [34].

In certain supergravity theories there existed an improved formation obtained by modifying the initial gauge invariant Lagrangian by addition of new fields. This Lagrangian has a gauge group defined without the equations of motion, as such it allows for gauge fixing procedure and ends up with an off-shell BRST symmetry. It is conceivable that such a formulation should be constructed in string field theory too.

The issue of precise Feynman rules in string field theory is of relevance when one considers loop diagrams. It was well known in dual model theory that at the loop level closed string states necessarily appear (as poles of the twisted one loop diagram). Consequently if the covariant field theory of open strings is complete one should have closed strings in the loop diagram. Arguments for this were given based on theorems on triangulation of modular space in Ref. [35], a more detailed explicit calculation was performed recently to exhibit the effect. One of the rather interesting still unsolved problems is, however, the following. In the light–cone gauge the closed string effect has to be explicitly added and is represented by an additional closed string field. It couples to the open string field and has a kinetic term (and also self–interacting terms) in the light–cone Lagrangian. Since in the covariant (Siegel gauge) open string Lagrangian nothing like that happens, the question is how these explicit closed string terms are introduced when changing from one gauge to another. Obviously this issue again calls for better understanding of the gauge fixing procedure in Witten's theory.

Concerning generalizations of the theory, one might also wonder about the relationship of Witten's theory and the covariant light–cone–type theory (it is described here by Dr. Hata). The possibility exists that one has a more general framework where the two theories are reached by partial gauge fixing. Proposals in this direction were made [24]. The basic ideas are appealing.

The most important topic of present investigations concerns closed strings. It is fair to say that the open string field theory as outlined by Witten is very well founded and is quite successful. Closed string field theory, however, does not quite exist yet. There are some interesting basic ideas. Proposals for analogous three string interactions were made but this approach is not complete and one might wonder whether it is impossible. What one has at this point is the fact that Witten's theory indeed contains closed strings (in loops) with no further modification. So it is a complete theory of open and closed strings and can serve as a starting point for a theory of purely closed strings.

An interesting aspect of string field theory is that an equivalent action, without background dependence, does exist [37-40]. The cubic action consists of just the interaction term of Witten's action

$$S = \frac{1}{3} \int A * A * A \ . \tag{123}$$

Shifting the field A by a solution of the equations of motion

$$A_0 * A_0 = 0 \ , \tag{124}$$

we get back Witten's action provided that the specific solution A_0 also satisfies

$$A_0 * A - (-)^A A * A_0 = QA \ . \tag{125}$$

It has been shown that

$$A_0 = (Q_L \ I)_B \ , \tag{126}$$

where Q_L is the integral of the BRST current over the "left piece" of the string, i.e., from 0 to $\frac{\pi}{2}$, and I is the identity element of the $*$ algebra. B denotes the explicit dependence of A_0 on the background in which we calculate the BRST current. As such, A_0 is the string field representing the background B. The proof of this comes from using the identities $QI = 0, \{Q, Q_L\} = 0$ and $Q_R A * B + (-)^A A * Q_L B = 0$. The first two equations are well known; the third follows from the overlap equations on the vertex. A verification was given using the explicit vertex operator. In all of these issues regularization plays an important role.

Let us note that the cubic action having no explicit metric dependence could possibly live in a space–time of arbitrary dimension. We learn that the space–time is picked by a particular solution of the equation of motion. Along this line, it has been argued that for a suitably chosen space of string functionals the cubic action may in fact be independent of the topology of space time. Again the topology of space–time would then be a property of the vacuum around which we expand.

The gauge invariance of the cubic action is

$$\delta_\lambda A = [A, \lambda] \equiv A * \lambda - \lambda * A \ , \tag{127}$$

where λ is an arbitrary string field of ghost number $-\frac{3}{2}$. Upon shifting to Witten's theory, the gauge invariance goes into the usual inhomogeneous one. Let us note in passing that in gauge fixing the cubic action we see the same problem of zero modes that exist only on-shell, as in Witten's theory. The precise off-shell meaning of both the Witten and cubic action is still not known.

To exhibit a closed string state in the open string theory, one proceeds as follows [41-43]. Introduce (linearized) gravitons by perturbing our flat metric by $\delta g_{\mu\nu} = E_{\mu\nu}\, e^{ik\cdot X}$ where $g^{\mu\nu} E_{\mu\nu} = k^\mu\, E_{\mu\nu} = k^2 = 0$, i.e., the graviton is "on-shell." Since we have obtained an expression for the background in terms of string fields, then the one (on-shell) graviton state is given by

$$h = \delta_h(Q_L I) \tag{128}$$

$$(\delta_h \equiv \delta g_{\mu\nu}).$$

Varying the action we find

$$\delta_h S = \frac{1}{2}\int (A * h * A + A * A * h). \tag{129}$$

This is just the coupling of the graviton to open string states. In the first quantized form this corresponds to

$$\delta\frac{1}{2}\int d^2\sigma \partial_\alpha\, X^\mu\, \partial^\alpha\, X^v \cdot G_{\mu\nu}(X)$$

$$= \frac{1}{2}\int d^2\sigma\, \partial_\alpha\, X^\mu \cdot \partial^\alpha\, X^v\, \delta g_{\mu\nu} = \int d^2\sigma V_h\ ,$$

where V_h is the graviton vertex operator. h can be explicitly calculated and a nontrivial regularization gives [41,43]

$$h = \Big(\int_0^{\pi/2} d\sigma^\alpha\, \epsilon_{\alpha\beta}\, C^\beta\, V_h\Big) I\ . \tag{130}$$

Similarly we can represent other on-shell closed string states.

At this point one can ask a question concerning the interpretation of this closed string state. Strominger has presented arguments that it is actually a solution to the appropriately defined $A * A = 0$ equations. We shall discuss this at the end of the section.

One clear interpretation of the closed string state is obviously the following. We know that it certainly appears as a pole in the one-loop diagram of Witten's open string theory. We should then consider the one-loop effective action

$$S_{eff} = \int (\hat{A} * Q\hat{A} + \frac{2}{3}\hat{A}^3) - \frac{1}{2}Tr\Big((Q+A)^2\Big)\ . \tag{131}$$

Looking at the second order term (in A) we can write it as

$$S_2 \sim \hat{A}\hat{T}\Delta_c\hat{T}\hat{A}\ , \tag{132}$$

where $\hat{T}$ stands for the open to closed string vertex operator found in the factorization of the pole and Δ_c is the closed string propagator. The <u>effective</u> equations of motion then become (in the leading order)

$$QA = \hat{T}C$$
$$Q_c C = 0, \tag{133}$$

where C is an explicit closed string field which is used to replace the pole term by the effective open-closed interaction $\hat{A}TC$. Clearly the solution to these equations is $C = V_h$ and

$$QA_{c\ell} = V_h \hat{T} \ . \tag{134}$$

One has an agreement with the earlier form since $V_h\hat{T}$ is essentially equal to Qh.

A much more profound statement is that the closed string state is actually a solution [41] of the classical cubic theory equations of motion. For this, one has to give a precise definition of the $*$ operator which would be applicable to states of the closed string form.

In string field theory, presently the single biggest gap may be not knowing precisely the space of string functionals in which the most general theory must live. The representation of the vertex (flat space–time) in oscillator representation applies to Fock space, however, we have important objects (for instance, the vertex itself) which have infinite norm and thus strictly lie outside of the corresponding Fock space.

If we look at string fields outside of the Fock space, then one has objects that do not satisfy Witten's axiom of associativity. (The order in which we do summation of two sums that are not absolutely convergent is important.) In fact, there is a difference between $(A*)B$ and $A(*B)$.

$$(A*)B = (\langle V_{123} \mid A_1 \rangle) \mid A_2 \rangle$$

$$A(*B) = (\langle V_{123} \mid B_2 \rangle) \mid A_1 \rangle .$$

If we restrict ourselves to the Fock space, Witten's axiom is OK. This space is then, however, not big enough to hold h. Moreover, in doing general nonperturbative shifts, we want our string space large enough so we don't get "thrown out." A working hypothesis is that we should consider all string fields that generate an inner derivation of the string field product. As an example: The hAA coupling does not violate Ward identities even though $Qh \neq 0$ because

$$-Q(h \cdot A) = h \cdot QA \ , \tag{135}$$

where $h \cdot A \equiv \mathcal{L}_h(A)$ is the inner derivative. Since we do not have associativity in general, the general open string cubic action could be defined as

$$S_+ = \frac{1}{6} \int (((A*)A*)\, A) + (A\, (*A(*A))) \ . \tag{136}$$

On the other hand, the choice

$$S_- = \frac{1}{6} \int (((A*)A*)\, A) - (A\, (*A(*A))) \tag{137}$$

vanishes on all open string states. As such, it is an excellent candidate for a purely closed string theory. Note that A is still an element of the space of open string fields. The integration in S_- in order to give a non–zero result must be performed over an appropriate closed string surface. Strominger has shown that this action gives the correct 3–closed–string coupling. The interesting question of higher point amplitudes is still open. The classical equations of motion are solved by $A_0 = (Q_L I)_B$.

ACKNOWLEDGEMENTS

Work supported by the Department of Energy, Contract DE-AC02-76ER03130.A026 - Task A.

REFERENCES

1. M. Kaku and K. Kikkawa, *Phys. Rev.* **D10** (1974) 1110,1823.
2. E. Cremmer and J.-L. Gervais, *Nucl. Phys.* **B76** (1979) 209; **B90** (1975) 410.
3. M. B. Green and J. M. Schwarz, *Nucl. Phys.* **B243** (1984) 475.
4. S. Mandelstam, *Nucl. Phys.* **B64** (1973) 205; **B69** (1974) 77, 83 (1974) 413.
5. J.-L. Gervais and B. Sakita, *Phys. Rev. Lett.* **30** (1973) 719.
6. A. M. Polyakov, *Phys. Lett.* **B103** (1981) 207,211.
7. D. Friedan, *Les Houches 1982*, B. Zuber and R. Stora, Eds., North-Holland 1984.
8. W. Siegel, *Phys. Lett.* **149B** (1984) 157,162; **151B** (1985) 391,396.
9. K. Fujikawa, *Phys. Rev.* **D25** (1982) 2582.
10. M. Kato and K. Ogawa, *Nucl. Phys.* **B212** (1983) 443.
11. W. Siegel and B. Zweibach, *Nucl. Phys.* **B263** (1986) 105.
12. T. Banks and M. Peskin, *Nucl. Phys.* **B264** (1986) 513.
13. K. Itoh, T. Kugo, M. Kunitomo and H. Ooguri, *Prog. Theor. Phys.* **75** (1986) 162.
14. A. Neveu, M. Nicolai and P. West, *Nucl. Phys.* **B264** (1986) 573.
15. E. Witten, *Nucl. Phys.* **B268** (1986) 253, *Nucl. Phys.* **B276** (1986) 291.
16. S. Giddings, *Nucl. Phys.* **B278** (1986) 242; S. Giddings and E. Martinec, *Nucl. Phys.* **B278** (1986) 91.
17. D. J. Gross and A. Jevicki, *Nucl. Phys.* **B283** (1987) Part I., *Nucl. Phys.* **B287** (1987) 225 Part II.
18. D. J. Gross and A. Jevicki, *"Operator Formulation of Interacting String Field Theory III: Superstrings,"* *Nucl. Phys.* **B293** (1987) 29.
19. A. Jevicki, *Fizika* **18, 36** (1986) - A Contribution to the Paul Senjanović Memorial Volume; *"Lectures on Interacting String Field Theory"* in the Proceedings of the Fifth Adriatic Meeting on Particle Physics, ed. by M. Martinis and I. Andrić, World Scientific (1987).
20. E. Cremmer, A. Schwimmer and C. Thorn, *Phys. Lett.* **179B** (1986) 57.
21. S. Samuel, *Phys. Lett.* **181B** (1986) 255; CCNY-HEP-8712.
22. K. Itoh, K. Ogawa and K. Suehiro *Nucl. Phys. B* (1987).
23. K. Suehiro, *Nucl. Phys.* **B296** (1988) 333;, *Prog. Theor. Phys.* **78** (1987) 1151.
24. M. Kaku *Intl. Journ. of Mod. Phys. A* Vol 2, No. 1 (1987) 1; CCNY preprint July (1987).
25. W. Siegel, *Phys. Lett.* **149B** (1984) 391.
26. H. Hata, K. Itoh, T. Kugo, H. Kunitomo and K. Ogawa, *Phys. Lett.* **172B** (1986) 186, *Phys. Rev.* **D34** (1986) 2360, D35 (1987) 1318.
27. A. Neveu and P. West, *Phys. Lett.* **168B** (1986) 192; A. Neveu and P. West, *Nucl. Phys.* **B268** (1986) 125; A. Neveu and P. West, *Nucl. Phys.* **B278** (1986) 601.
28. A. Jevicki, *Phys. Lett.* **169B** (1986) 359.
29. A. Jevicki and B. Sazdović, *"Supersymmetry in the Operator Formulation of Witten's Field Theory"*, Brown preprint HET-615 (1987).

30. T. Kugo and H. Terao, Kyoto University preprint, KUNS 911.
I. Ya. Aref'eva and P. B. Medvedev, Steklov Math. Inst. preprint (1987).

31. M. Bocchichio, *Phys. Lett.* **B188** (1987) 330, *Phys. Lett.* **182B** (1987) and Princeton preprints (1987).

32. C. Thorn, *Nucl. Phys.* **B287** (1987) 61.

33. A. Bogojević, *Phys. Lett.* **B198** (1987) 479.

34. Č. Crnković and E. Witten, *"Covariant Description of Canonical Formalism in Geometrical Theories"*, Princeton preprint (1986); Č. Crnković, Princeton preprint (1987).

35. S. Giddings, E. Martinec and E. Witten, *Phys. Lett.* **176B** (1986) 362.

36. J. A. Shapiro and C. Thorn, IAS-HEP-87/20; IAS-HEP-8717 (1987);
D. Freedman, S. Giddings, J. A. Shapiro and C. Thorn, IAS preprint (1987).

37. G. T. Horowitz, J. Lykken, R. Rohm and A. Strominger, *Phys. Rev. Lett.* **57** (1986) 283.

38. H. Hata, K. Itoh, T. Kugo, H. Kunitomo and K. Ogawa, *Phys. Lett.* **175B** (1986) 138.

39. D. Friedan, *Phys. Lett.* **162B** (1985) 102.

40. T. Yoneya, in *Proceedings of the Seventh Workshop on Grand Unification*, Toyama, Japan (1985).

41. A. Strominger, *Phys. Rev. Lett.* **58** (1987) 629, *Phys. Lett.* **B187** (1987) 295.

42. G. Horowitz and A. Strominger, *Phys. Lett.* **B185** (1987) 45.

43. M. Srednicki and R. Woodard, *Nucl. Phys. B* (to appear).

44. C. Wendt, SLAC-Pub-4442 (1987).

45. M. Green and N. Seiberg, IAS preprint (1987).

Gauge String Field Theory*

H. Hata

Department of Physics, Kyoto University, Kyoto 606, Japan

Abstract: A brief survey of gauge string field theory is given.

There are a number of approaches to (super)string theory. Among them the most popular one would be the 1st quantized approach initiated by Polyakov.[1] There have been many talks in this direction presented in this Symposium. Here, however, I would like to talk about another approach: It is the *string field theory*.

Before beginning the discussion, let me summarize why we should consider string field theory. In my opinion there are mainly three reasons: The first is that field theory is the most conventional and well-developed framework to study a quantum (or classical) system. It is not restricted merely to S-matrix or perturbation theory. For example, let me pose the following two questions in string theory:

i) What is the stable vacuum in string theory?

ii) Do we have Black Holes in string theory?

In the framework of string field theory we could, at least in principle, answer these questions. For the first question we have a very powerful technique of effective potential. The second question could be answered by studying the (classical) equation of motion in string field theory. However, it is not straightforward to answer these questions on the basis of the 1st quantized approach.

The second reason for favoring string field theory is that the underlying gauge principle in string theory is manifest there. As is well known, string theory is a *local gauge theory* (here I mean the local gauge symmetry in *space-time* sense but not in the 2-dimensional world-sheet sense), and it is a kind of extension of Yang-Mills theory or Einstein theory. In fact, in the zero-slope limit $\alpha' \to 0$, string theory is known to reduce to these familiar local gauge theories.[2,3] Therefore, string theory should also be formulated in such a way that its (space-time) gauge symmetry is manifest. This requirement leads us naturally to string field theory. (On the other

⋆ Work supported in part by Grant-in-Aid for Scientific Research from the Ministry of Education, Science and Culture (#62740153).

hand, space-time gauge symmetry, or space-time concepts in general, are very indirect in the 1st quantized formalism.)

Finally, the third reason for considering string field theory is that it is possible, in the framework of string field theory, to construct a formulation of string theory which does not depend on a particular space-time background. This will be discussed in the final part of my talk.

Having these motivations in mind, let us turn to the construction of string field theory. Before that I should make a comment: In this talk I will often sacrifice mathematical preciseness for the sake of simplicity. Therefore interested readers should consult the original references for the precise expressions of various equations.

What we search for is now evident. It is a stringy extension of Yang-Mills action or Einstein action and their associated local gauge symmetry transformations:

$$\begin{aligned} S_{\text{YANG-MILLS}} &= -\frac{1}{4g^2}\int dx\, F_{\mu\nu}^2 \quad ; \quad \delta A_\mu = \partial_\mu \lambda + A_\mu \times \lambda \\ S_{\text{EINSTEIN}} &= \frac{1}{\kappa^2}\int dx\, \sqrt{-g}R \quad ; \quad \delta g_{\mu\nu} = \xi_{\mu;\nu} + \xi_{\nu;\mu}\,. \end{aligned} \tag{1}$$

At present there are two formulations of string field theory: One is Witten's fomulation[4] and this was discussed in detail by Prof. Jevicki in this Symposium. In Witten's string field theory one can treat only the open string theory. The other formulation of string field theory was presented by Kyoto group, namely by Itoh, Kugo, Kunitomo, Ogawa and myself.[5,6,7] (A similar formulation was given partly by Neveu and West.[8]) Contrary to Witten's one, Kyoto group formalism can treat both open and closed string theories.

Although these two formulations of string field theory differ in details, their gauge invariant action and the associated gauge transformation take in both cases the following form†

$$S = \Phi \cdot Q_{\text{B}} \Phi + \frac{2}{3} g \Phi \cdot (\Phi * \Phi) \tag{2}$$

$$\delta\Phi = Q_{\text{B}}\Lambda + 2g\Phi * \Lambda\,. \tag{3}$$

Now, let me explain the basic elements which appear in (2) and (3). These are string field Φ, BRS operator Q_{B}, star-product $*$ and dot-product $\cdot$.

a) String field Φ: The string field Φ is a functional of string coordinate $X_\mu(\sigma)$ as well as the (fermionic) Faddeev-Popov (FP) ghost coordinates $c(\sigma)$ and $\bar{c}(\sigma)$:

$$\Phi = \Phi\left[X_\mu(\sigma), c(\sigma), \bar{c}(\sigma)(, \alpha)\right]. \tag{4}$$

† Precisely speaking, we need to add a Φ^4 term and a $\Phi^2\Lambda$ term to the action (2) and the gauge transformation (3), respectively, in the case of Kyoto group open string field theory. Although we neglect them for simplicity, interested readers should consult Ref. 6 for details.

In the case of Kyoto group formalism the string field Φ depends also on another (real) variable α which we call the string-length parameter. The necessity of this variable α will be explained when we construct the star-product.

b) BRS operator Q_{B}: This is a linear operator which acts on the string fields (string functionals).

We have two kinds of product between two string fields (functionals) in (2) and (3):

c)Star-product $\Phi * \Psi$: The star-product of two string functionals Φ and Ψ gives another string functional $\Phi * \Psi$.*

d) Dot-product $\Phi \cdot \Psi$: This product gives a number.

Therefore, if we compare the string functional Φ to a (3-dimensinal) vector $\boldsymbol{a}$, then the following correspondence holds:

$$\begin{aligned} \Phi, \Psi &\leftrightarrow \boldsymbol{a}, \boldsymbol{b} \quad \text{(vector)} \\ \Phi * \Psi &\leftrightarrow \boldsymbol{a} \times \boldsymbol{b} \quad \text{(vector product)} \\ \Phi \cdot \Psi &\leftrightarrow \boldsymbol{a} \cdot \boldsymbol{b} \quad \text{(scalar product)}\ . \end{aligned} \tag{5}$$

Finally, Λ appearing in (3) is the parameter of gauge transformation and it is also a string functional.

Of course, the above basic elements, in particular, Q_{B}, star-product and dot-product, cannot be arbitrary but they have to satisfy certain requirements in order for the gauge invariant action (2) to be really invariant under the gauge transformation (3). These requirements on the set $\left(Q_{\mathrm{B}}, *, \cdot\right)$ are as follows:‡

$$\begin{aligned} &i)\ (Q_{\mathrm{B}})^2 = 0 \quad \text{(nilpotency)} \\ &ii)\ Q_{\mathrm{B}}(\Phi * \Psi) = (Q_{\mathrm{B}}\Phi) * \Psi + \Phi * (Q_{\mathrm{B}}\Psi) \quad \text{(Leibniz rule)} \\ &iii)\ (\Phi * \Psi) * \Sigma + (\Psi * \Sigma) * \Phi + (\Sigma * \Phi) * \Psi = 0 \quad \text{(Jacobi identity)} \\ &iv)\ (Q_{\mathrm{B}}\Phi) \cdot \Psi = -\Phi \cdot (Q_{\mathrm{B}}\Psi) \quad \text{(partial differentiability of } Q_{\mathrm{B}}) \\ &v)\ \Phi * \Psi = -\Psi * \Phi \\ &vi)\ \Phi \cdot (\Psi * \Sigma) = \Psi \cdot (\Sigma * \Phi) \quad \text{(cyclic symmetry)} \end{aligned} \tag{6}$$

In *ii)-vi)*, Φ,Ψ and Σ are arbitrary string functionals. Extending the analogy (5), the requirements *iii), v)* and *vi)* have the following correspondence with the familiar properties of the (3-dimensional) vectors:

$$\begin{aligned} iii) &\leftrightarrow (\boldsymbol{a} \times \boldsymbol{b}) \times \boldsymbol{c} + (\boldsymbol{b} \times \boldsymbol{c}) \times \boldsymbol{a} + (\boldsymbol{c} \times \boldsymbol{a}) \times \boldsymbol{b} = 0 \\ v) &\leftrightarrow \boldsymbol{a} \times \boldsymbol{b} = -\boldsymbol{b} \times \boldsymbol{a} \\ vi) &\leftrightarrow \boldsymbol{a} \cdot (\boldsymbol{b} \times \boldsymbol{c}) = \boldsymbol{b} \cdot (\boldsymbol{c} \times \boldsymbol{a}) \end{aligned} \tag{7}$$

---

* In the open string case the star-product $*$ here is related to the original star-product $\star$ defined in Refs. 4 and 6 by $\Phi * \Psi \equiv (1/2)\{\Phi \star \Psi - (-)^{|\Phi||\Psi|}\Psi \star \Phi\}$. When the $\star$ product satisfies the associativity, the present $*$ product satisfies the Jacobi identity.

‡ Strictly speaking, we have to attach certain sign factors to each term in (6) depending on the Grassmann even-oddness of the string functionals. See Refs. 6 and 7.

One can easily show the gauge invariance of the action (2), $\delta S = 0$, by using the above properties $i) - vi)$. Since this holds irrespectively of the concrete choice of $(Q_B, *, \cdot)$ so long as they satisfy (6) , there is a possibility that we can construct many different string field theories (although their physical contents might be the same). In fact, two formulations of string field theory we have at present, *i.e.*, Witten's and Kyoto group one, differ in their choice of star-product.

Therefore, the construction of gauge invariant string field theory has been reduced to that of $(Q_B, *, \cdot)$ which satisfy (6). First, as the BRS operator Q_B, let us take the BRS charge which appears in the covariant 1st quantization of string presented by Kato and Ogawa.[9] It takes the following form:

$$\begin{aligned} Q_B =& \frac{\sqrt{\pi}}{4}\sum_{\pm}\int d\sigma \left(\frac{\delta}{\delta\bar{c}(\sigma)} \pm c(\sigma)\right)\left[-\left(-i\frac{\delta}{\delta X^{\mu}(\sigma)} \pm X^{\mu\prime}(\sigma)\right)^2\right. \\ & \left. \pm 2i\frac{d}{d\sigma}\left(\frac{\delta}{\delta\bar{c}(\sigma)} \pm c(\sigma)\right)\cdot\left(\bar{c}(\sigma) \pm \frac{\delta}{\delta c(\sigma)}\right)\right] \\ =& \sum_{n=-\infty}^{\infty} :c_{-n}\left(L_n^X + \frac{1}{2}L_n^{\mathrm{FP}} - \alpha(0)\delta_{n,0}\right): \quad , \end{aligned} \tag{8}$$

where L_n^X and L_n^{FP} are the Virasoro operators constructed out of $X_\mu(\sigma)$ and FP ghost coordinates $c(\sigma)$ and $\bar{c}(\sigma)$, respectively. As shown by Kato and Ogawa,[9] the nilpotency of this BRS operator Q_B holds only when the space-time dimension $D = 26$ and the zero-intercept parameter $\alpha(0) = 1$.

Next is the dot-product which gives a number out of two string functionals. We define it as follows:

$$\Phi\cdot\Psi \equiv \int \mathcal{D}Z(\sigma)\,\Phi\left[Z(\sigma)\right]\Psi\left[Z(\sigma)\right] \ , \tag{9}$$

where $Z(\sigma)$ is the set of string coordinates

$$Z(\sigma) \equiv \left(X_\mu(\sigma), c(\sigma), \bar{c}(\sigma)\,(,\alpha)\right) \ . \tag{10}$$

Equation (9) is a natural stringy extention of the scalar product for local functions,

$$\int dx\, \phi(x)\psi(x) \ .$$

With this definition of dot-product (9) the partial differentiability of the BRS operator, *iv)* of (6) , is seen to hold.

Finally let us consider the star-product. This is the most difficult and important thing, and it determines the form of the string interactions in string field theory. The star-product gives another string functional from two string functionals, and we assume the following form for it:

$$(\Phi * \Psi)[Z_3] \equiv \int\int \mathcal{D}Z_1\mathcal{D}Z_2 V[Z_1, Z_2, Z_3]\Phi[Z_1]\Psi[Z_2] \tag{11}$$

with $V[Z_1, Z_2, Z_3]$ being a certain functional of three string coordinates Z_{1-3}. The star-product must satisfy a number of requirements, in particular, Leibniz rule and Jacobi identity. First, from the requirement of Leibniz rule, *ii)* of (6), it follows that $V[Z_1, Z_2, Z_3]$ should essentially be a "delta-functional" of string coordinates $Z_i(\sigma_i)$ $(i = 1-3)$ expressing the connection *two* $Z_i(\sigma_i)$ at each string point σ. Examples of $V[Z_1, Z_2, Z_3]$ having this property are given in Fig. 1 for the open string case and in Fig. 2 for the closed string case. These figures mean that $V[Z_1, Z_2, Z_3]$ is a delta-functional of string coordinates which expresses the connection indicated by them.

First, for the open string, Fig. 1(A) is the string connection adopted by Witten,[4] and this has been explained in detail by Prof. Jevicki. Here, the three strings coincide at their mid-points $Z_i(\sigma = \frac{\pi}{2})$. On the other hand, Fig. 1(B) is the string connection in Kyoto group open string field theory.[5,6,7] It is of the same form as the string connection in the light-cone gauge string field theory of Kaku-Kikkawa.[10] However, if we are to adopt this type of string connection, we have to specify where the break point is. In the case of light-cone gauge string field theory the break point was determined by the light-cone momentum p^+. This is not allowed in our case because of Lorentz covariance, and we are forced to introduce the string-length parameter α. α is something momentum-like variable, which is conserved in $V[Z_1, Z_2, Z_3]$, and the break point is determined in the parameter space of σ as if each string has its length α. [α is not the real string length in space-time. It is the "length" in the σ-space.]

Similarly for the closed string field theory, two types of the string connection can be considerd: Figure 2(A) is the string connection discussed by Lykken-Laby[11] (and probably many others). This is a naive closed string extension of the Witten's open string connection of Fig. 1(A). Figure 2(B) is that of Kyoto group closed string field theory. The interaction point (*i.e.*, the pinched point) is determined by the string-length α.

For each of the string connections (*i.e.*, $V[Z_1, Z_2, Z_3]$'s) given in Figs. 1 and 2, we can show that the Leibniz rule holds when $D = 26$. However, the star-product has to satisfy another requirement, the Jacobi identity, and this imposes a severer restriction on the possible form of

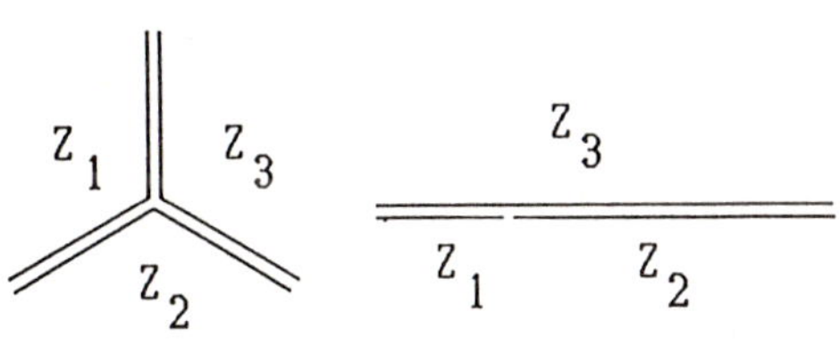

(A) **Figure 1** (B)

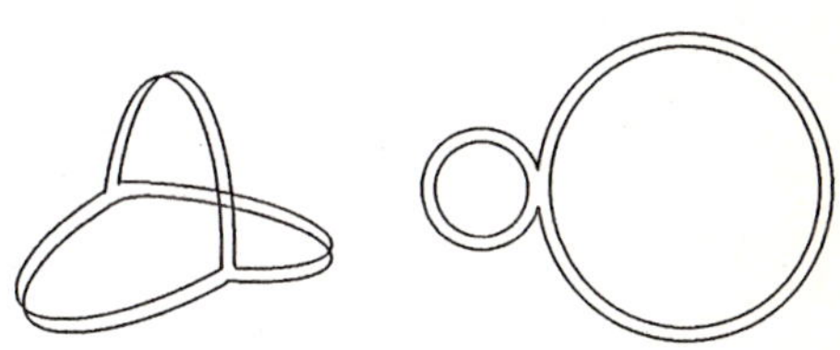

(A) **Figure 2** (B)

Figure 1. String connection diagrams for the star-product in open string field theory.

Figure 2. String connection diagrams for the star-product in closed string field theory.

the star-product: First, the star-products of Kyoto group string field theory given by Fig. 1(B) for the open string and by Fig. 2 (B) for the closed string are both shown to satisfy the Jacobi identity when $D = 26$.[5,6,7] However, the closed string star-product of Fig. 2(A) fails to satisfy the Jacobi identity and hence it is ruled out.[7] The star-product of Witten given by Fig. 1(A) is probably all right, but I know of no proof at present.

Figure 3 shows how the Jacobi identity is realized for the star-product of Kyoto group closed string field theory given by Fig. 2(B).[7] To each of the three terms on the LHS of the Jacobi identity, *iii)* of (6), corresponds a string connection diagram of the three strings Φ, Ψ and Σ (diagrams (a), (b) and (c)), and the Jacobi identity requires that these three diagrams cancel one another. [In diagram (a), the dotted string is the intermediate string $\Phi * \Psi$, and the final string $(\Phi * \Psi) * \Sigma$ is indicated by the broken line. The meaning of the other diagrams is quite similar.] At first sight, one might think that such a cancellation is impossible since the string connections of (a)-(c) are completely different. However, the string connections corresponding to the three terms on the LHS of the Jacobi identity are not only those given by (a)-(c), but they

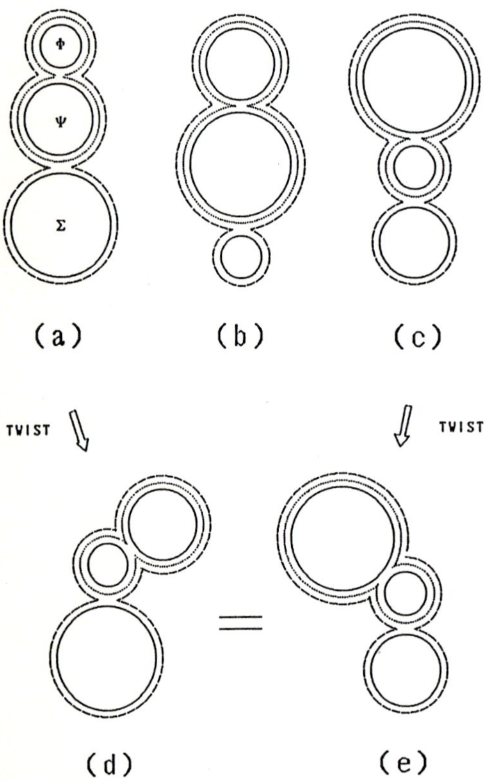

Figure 3. Illustration of how the Jacobi identity is realized in closed string field theory of Kyoto group. In each diagram the smallest string (solid line) represents Φ, the middle size one Ψ and the largest one Σ.

in fact contain infinitely many others obtained from (a)-(c) by twisting (rotating) the strings appearing there. This is because of the following constraint imposed on the closed string field,

$$\Phi_{closed}[Z(\sigma + \text{const.})] = \Phi_{closed}[Z(\sigma)] \ , \tag{12}$$

which implies that the origin of the σ-coordinate is irrelevant for the closed string. For example, by twisting the intermediate dotted string in (a) we obtain diagram (d). On the other hand, by the same procedure we obtain (e) from (c). As seen by turning diagram (e) upside down, (d) and (e) express the same connection for the strings Φ, Ψ and Σ, and they in fact cancel each other (when $D = 26$). Similar cancellations can be shown to occur totally to realize the Jacobi identity when $D = 26$.

As I explained above, the Kyoto group string field theory is described by the gauge invariant action (2) with the star-product given by Fig. 1(B) and Fig. 2(B). It expresses the string interaction of joining and splitting type shown in Fig. 4. Although we have omitted it from the discussion for the sake of simplicity, we have to add the 4-string interction term Φ^4 to the open string action, which corresponds to the exchange type interaction also shown in Fig. 4. [Strictly speaking, there is a breaking of the Jacobi identity in the Kyoto group open string field theory, and the Φ^4 term is added to correct this breaking. See Ref. 6 for details.]

Then, let us summarize the main properties of the Kyoto group string field theory.

1. Correct string scattering ampliutudes are obtained at the tree level for both the open and the closed string cases,[6,7] and also at the 1-loop level for the open string case.[12]

2. In the zero-slope limit $\alpha' \to 0$, our open string field theory action (2) and the gauge transformation (3) are shown to reproduce those of the Yang-Mills theory with an additional continuous "index" $\widetilde{\alpha}$ (which is the variable dual to α):[6]

$$\begin{aligned} S_{\text{YANG-MILLS}} &= -\frac{1}{4g_{\text{YM}}^2}\int dx\, d\widetilde{\alpha}\, F_{\mu\nu}^2(x,\widetilde{\alpha}) \\ \delta A_\mu(x,\widetilde{\alpha}) &= \partial_\mu\lambda(x,\widetilde{\alpha}) + A_\mu(x,\widetilde{\alpha})\times\lambda(x,\widetilde{\alpha}) \ , \end{aligned} \tag{13}$$

where $A_\mu(x,\widetilde{\alpha})$ and $F_{\mu\nu}(x,\widetilde{\alpha})$ are defined by

$$\begin{aligned} A_\mu(x,\widetilde{\alpha}) &\equiv \int_{-\infty}^{\infty}\frac{d\alpha}{2\pi}e^{i\alpha\widetilde{\alpha}}A_\mu(x,\widetilde{\alpha}) \\ F_{\mu\nu}(x,\widetilde{\alpha}) &\equiv \partial_\mu A_\nu(x,\widetilde{\alpha}) - \partial_\nu A_\mu(x,\widetilde{\alpha}) + A_\mu(x,\widetilde{\alpha})\times A_\nu(x,\widetilde{\alpha}) \end{aligned} \tag{14}$$

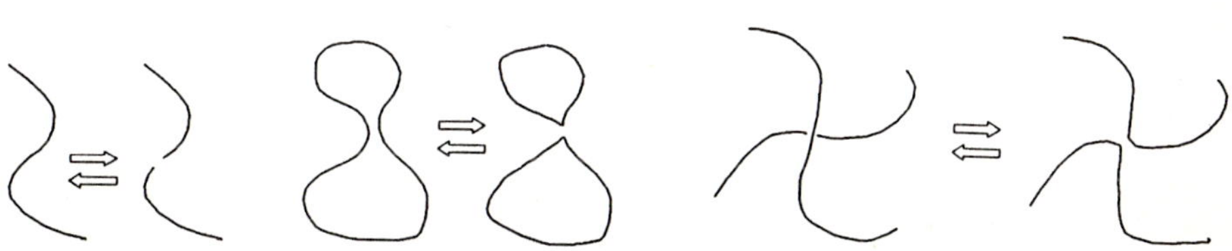

Figure 4. The string interactions in Kyoto group string field theories.

and $A_\mu(x,\alpha)$ is the Yang-Mills local field which appears in the mode expansion of the string field $\Phi\left[X_\mu(\sigma), c(\sigma), \bar{c}(\sigma), \alpha\right]$.

3. Our string field contains string-length parameter α as its variable, and one might be afraid that this causes problems: For example, we have an infinite degeneracy of string states. There arise additional divergeces in muti-loop diagrams due to the loop α-integrations.

However, these are, at least formally, not problems! First, we can show that the physical scattering amplitudes are independent of the external state string-length α_i aside from the conservation delta-function $\delta(\sum_i \alpha_i)$. This implies that the physical state vector spaces with different $\widetilde{\alpha}$, which is the dual variable to α (cf. (14)), are *equivalent* but completely *disconnected.* Therefore, the infinite degeneracy problem is resolved by considering a single physical state vector spaces with some definite $\widetilde{\alpha}$.[12] As for the additional divergences in multi-loop diagrams, they are, at least formally, shown to be absorbed into the coupling constant g or $\hbar$.[12]

4. In our (pure) open string field theory, the closed string states appear as the *bound states*.[12] This is a very surprising property in view of the situation in the light-cone gauge string field theory[10] (where the pure open string system is inconsistent and the elementary closed string field must be introduced by hand). As discussed in Prof. Jevicki's talk, this property is shared also by Witten's open string field theory.[4,13]

5. The last is a rather disappointing and serious problem in closed string field theory. If we calculate the 1-loop amplitude in our closed string field theory, the resultant amplitude covers the fundamental region infinitely many times and hence the unitarity is violated.[12] At present, we have no definite idea to remedy this defect of closed string field theory. (We suspect that this is related to the problem of quantization or gauge-fixing.) Anyhow, we hope to resolve this problem in the near future.

Finally in my talk, let me discuss a more ambitious approach in string field theory, *i.e.*, pregeometrical string field theory.[14]

Consider the closed string field theory of Kyoto group described by the gauge-invariant action (2) with the star-product defined by Fig. 2(B). As is well known, closed string theory contains graviton and it is an extension of Einstein gravity theory. However, even at the classical level the action (2) has an unpleasant feature that it depends explicitly on the flat space-time metric $\eta_{\mu\nu} = (-,+,+,\cdots,+)$. In fact, the BRS operator Q_B in the kinetic term $\Phi \cdot Q_\text{B}\Phi$ contains the following terms:

$$Q_\text{B} \sim \int d\sigma \left\{ \frac{\delta}{\delta c} \left(\eta^{\mu\nu} \frac{\delta}{\delta X^\mu} \frac{\delta}{\delta X^\nu} + \cdots \right) \cdots \right\} . \tag{15}$$

In addition, the graviton field $h_{\mu\nu}$ which appears in the mode expansion of the string field Φ in the form $|\Phi\rangle \sim h_{\mu\nu} \alpha^\mu_{-1} \widetilde{\alpha}^\nu_{-1} |0\rangle$ is not the total $g_{\mu\nu}$ but is a fluctuation on the flat metric $\eta_{\mu\nu}$; $g_{\mu\nu} \sim \eta_{\mu\nu} + h_{\mu\nu}$. Therefore, the construction of the theory is essentially perturbation theoretical upon the flat space-time background. Then a question naturally arises whether it is possible to reformulate the closed string field theory given by (2) into a form which is independent of a particular space-time background.

This is indeed possible. The basic idea is as follows: Although the kinetic term $\Phi\cdot Q_B\Phi$ of (2) is background-dependent, its interaction term $\Phi\cdot(\Phi*\Phi)$ is independent of the background since it merely expresses the connection of three strings. This idea was first suggested by Friedan[15] and Yoneya.[16] Therefore, let us consider a theory described by the following action which consists solely of the interaction term:

$$S_{pre} = \frac{2}{3g^2}\Psi\cdot(\Psi*\Psi)\,, \tag{16}$$

which has an invariance under the homogeneous local gauge transformation:

$$\delta_{pre}\Psi = 2\Psi*\Lambda\,. \tag{17}$$

This is the *pregeometrical string field theory* proposed by Kyoto group.[14] (We often abbreviate it simply to pre-theory hereafter.) A similar string field theory with Φ^3 action based on Witten's open string field theory was later developed by Horowitz *et al.*[17]

Although the preaction (16) is background independent, it lacks the kinetic term, without which the strings cannot propagate. Then, how the kinetic term can be generated in this pre-theory? The answer is that the kinetic term is generated through the *condensation* of the prefield Ψ in (16). Suppose that the prefield Ψ develops a non-vanishing vacuum expectation value $\langle\Psi\rangle = \Psi_0$. This Ψ_0 should satisfy the equation of motion of pre-theory,

$$\Psi_0*\Psi_0 = 0\,. \tag{18}$$

Then expressing the prefield Ψ as Ψ_0 plus the fluctuation Φ,

$$\Psi = \Psi_0 + g\Phi\,, \tag{19}$$

the preaction (16) and the pre-gauge-transformation (17) are expanded as

$$S_{pre} = \Phi\cdot Q\Phi + \frac{2}{3}g\Phi\cdot(\Phi*\Phi) \tag{20}$$

$$\delta_{pre}\Phi = Q\Lambda + 2g\Phi*\Lambda\,, \tag{21}$$

where Q is a linear operator defined by

$$Q\Phi \equiv 2\Psi_0*\Phi\,. \tag{22}$$

Equations (20) and (21) are of the same form as the previous (2) and (3), respectively, except that Q_B is replaced by Q above. Although this operator Q (22) is in general different from the Q_B of

(8), it should be noted that Q satisfies the same properties as Q_B does; *i.e.*, nilpotency, Leibniz rule and the partial differentiability (*i), ii)* and *iv)* of (6)). The nilpotency is a consequence of the Jacobi identity and the equation of motion (18):

$$Q^2\Phi = 4\Psi_0 * (\Psi_0 * \Phi) \propto \Phi * (\Psi_0 * \Psi_0) = 0\,.$$

Similarly, the Leibniz rule and the partial differentiability can be shown to hold by making use of the Jacobi identity and the cyclic symmetry (*vi)* of (6)), respectively. Therefore, the operator Q is regarded as a generalization of the previous BRS operator Q_B on the flat background.

Now, our interpretation is as follows: To each (classical) solution Ψ_0 of the equation of motion (18) of pre-theory corresponds an allowed space-time background, and the operator Q (22) is nothing but the BRS operator on this (possibly curved) background.

Then, is it actually possible to construct Ψ_0 which corresponds to, for example, flat backgound? The answer is Yes. First, let me give an intuitive explanation. Suppose that the α-dependence of Ψ_0 is the delta-function $\delta(\alpha)$. Namely, consider the condensation of infinitesimally "small" strings. The star-product of such Ψ_0 and a general string functional Φ is illustrated in Fig. 5. As seen from Fig. 5 the effect of the star-multiplication by Ψ_0 is an infinitesimal deformation along the string Φ. [Note that the small blob of Ψ_0 moves around Φ due to the constraint (12) on the closed string field.] And this is nothing but what the *kinetic operator* Q_B does on a string functional. Therefore, it is very likely that by choosing a suitable Ψ_0 with vanishing α, we can realize $\Psi_0 * \Phi \propto Q_B\Phi$.

In general, the following is known:[14] For any given BRS operator Q which satisfies both nilpotency and Leibniz rule, the corresponding classical solution Ψ_0 is given by

$$\Psi_0 = -\frac{1}{2}Q\Gamma, \tag{23}$$

where $\Gamma = \Gamma\,[Z(\sigma)]$ is a string functional which has the property

$$\Gamma * \Phi = \left(N_{FP} + 1 - \alpha\frac{\partial}{\partial\alpha}\right)\Phi \qquad \text{for } \forall\Phi \tag{24}$$

with N_{FP} being the ghost number operator. We can actually construct the string functional Γ which satisfies (24). [See Ref. 14 for details.] It can easily be shown that (18) and (22) hold for Ψ_0 given by (23) by using (24) and the nilpotency and the Leibniz rule of Q.

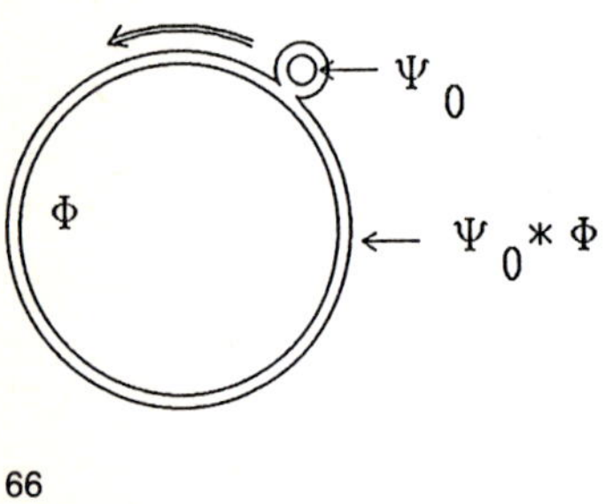

Figure 5. The string product $\Psi_0 * \Phi$. The string Ψ_0 has an infinitesimally small α and moves around Φ.

Now we have seen the one to one correspondence between the classical solution Ψ_0 of the equation of pre-theory, $\Psi_0 * \Psi_0 = 0$, and the BRS operator Q which satisfies nilpotency and Leibniz rule. In this connection, there are interesting papers by Banks-Nemeschansky-Sen[18] and by Kikkawa-Maeno-Sawada,[19] which show the connectioin between the present string field theory approach and the 1st quantized approach. First, from the requirement of the nilpotency of the BRS operator on a curved backgound Banks-Nemeschansky-Sen have derived the equation of motion for the backgound metric, anti-symmetric tensor and the dilaton, which is the same equation of motion as that obtained from the β-function approach. On the other hand, Kikkawa-Maeno-Sawada examined the Leibniz rule for the BRS operator on a curved background, and from it they obtained the same equation of motion as above.

It would be very interesting to study further the equation of motion of pre-theory, $\Psi_0 * \Psi_0 = 0$. The Black Hole in string theory might be contained in its solutions.

ACKNOWLEDGEMENTS

I would like to thank K. Itoh, T. Kugo, H. Kunitomo, K. Ogawa and K. Suehiro.

REFERENCES

1. A.M. Polyakov, *Phys. Lett.* **103B** (1981) 207; **103B** (1981) 211.
2. J. Scherk, *Nucl. Phys.* **B31** (1971) 222,
   A. Neveu and J. Scherk, *Nucl. Phys.* **B36** (1972) 155.
3. T. Yoneya, *Prog. Theor. Phys.* **51** (1974) 1907,
   J. Scherk and J.H. Schwarz, *Nucl. Phys.* **B81** (1974) 118.
4. E. Witten, *Nucl. Phys.* **B268** (1986) 2530; **B276** (1986) 291.
5. H. Hata, K. Itoh, T. Kugo, H. Kunitomo and K. Ogawa, *Phys. Lett.* **172B** (1986) 186; 195; *Nucl. Phys.* **283** (1987) 433.
6. H. Hata, K. Itoh, T. Kugo, H. Kunitomo and K. Ogawa, *Phys. Rev.* **D34** (1986) 2360.
7. H. Hata, K. Itoh, T. Kugo, H. Kunitomo and K. Ogawa, *Phys. Rev.* **D35** (1987) 1318.
8. A. Neveu and P.C. West, *Phys. Lett.* **168B** (1986) 192; *Nucl. Phys.* **B268** (1986) 125.
9. M. Kato and K. Ogawa, *Nucl. Phys.* **B212** (1983) 443.
10. M. Kaku and K. Kikkawa, *Phys. Rev.* **D10** (1974) 110; 1823,
    S. Mandelstam, *Nucl. Phys.* **B64** (1973) 205; **B69** (1974) 77; **B83** (1974) 413,
    E. Cremmer and J.-L. Gervais, *Nucl. Phys.* **B76** (1974) 209; **B90** (1975) 410.
11. J. Lykken and S. Raby, *Nucl. Phys.* **B278** (1986) 256.

12. H. Hata, K. Itoh, T. Kugo, H. Kunitomo and K. Ogawa, *Phys. Rev.* **D35** (1987) 1356.

13. S. Giddings, E. Martinec and E. Witten, *Phys. Lett.* **176B** (1986) 362.

14. H. Hata, K. Itoh, T. Kugo, H. Kunitomo and K. Ogawa, *Phys. Lett.* **175B** (1986) 362.

15. D. Friedan, *Nucl. Phys.* **B271** (1986) 540.

16. T. Yoneya, in *Proceedings of '7th Workshop on Grand Unification/ICOBAN'86.*

17. G.T. Horowitz, J. Lykken, R. Rohm and A. Strominger, *Phys. Rev. Lett.* **57** (1986) 283.

18. T. Banks, D. Nemeschansky and A. Sen, *Nucl. Phys.* **B277** (1986) 67.

19. K. Kikkawa, M. Maeno and S. Sawada, *Phys. Lett.* **197B** (1987) 524.

Part III

Loop Corrections and Conformal Field Theory

String Multi-loop Corrections to Equations of Motion

H. Ooguri[1] *and N. Sakai*[2]

[1] Department of Physics, University of Tokyo, Hongo, Bunkyo-ku, Tokyo 113, Japan

[2] Department of Physics, Tokyo Institute of Technology, Oh-okayama, Meguro-ku, Tokyo 152, Japan

The recently developed operator formalism on Riemann surfaces is used to study nonlinear sigma models on higher genus surfaces. Short-distance divergences due to small handles are evaluated with handle operators which implement the higher genus periodicity in the operator formalism. Self-energy corrections from multi-loop amplitudes are explicitly shown to exponentiate and to give loop corrections to anomalous dimensions. The Schottky parametrization is found to be useful in extracting the short-distance divergences.

1. Introduction

It is necessary to study the vacuum configurations of string in order to decide whether the superstring theory can explain the real physical world. The string propagation in background fields is described by a nonlinear sigma model. It has been found that the conformal invariance of the nonlinear sigma model on a two-sphere gives the equations of motion for the background fields at the string tree level [1-9]. Namely consistent vacuum configurations are determined by requiring the absence of short-distance divergences (no conformal anomaly). The standard argument supporting this idea is the following. The stability of the vacuum configuration requires that tadpole amplitudes vanish. A tree tadpole amplitude $T_{(0)}$ of a string state in the background field is given by

$$T_{(0)} = \frac{1}{V(SL(2, C))} \int d^2z \langle V(z) \rangle , \qquad (1.1)$$

Springer Proceedings in Physics Vol. 31: **Quantum String Theory**
Editors: N. Kawamoto and T. Kugo

where $\langle V(z)\rangle$ is an expectation value of a vertex operator for the string state in the corresponding nonlinear sigma model on the two-sphere, and V(SL(2, C)) is the volume of SL(2, C) given by

$$V(SL(2, C)) = \int d^2z_1 d^2z_2 d^2z_3 |z_1-z_2|^{-2} |z_2-z_3|^{-2} |z_3-z_1|^{-2}. \qquad (1.2)$$

Since V(SL(2, C)) is divergent in coincidence limits $z_i \to z_j$ (i, j=1, 2, 3), the tadpole amplitude (1.1) vanishes provided $\langle V(z)\rangle$ does not diverge. This absence of the short-distance divergence is assured by the conformal invariance. The leading order in the weak field expansion of the nonlinear sigma model is an insertion of a single vertex operator, whose conformal invariance leads to a linearized equation of motion for the background field. Nonlinear terms in the equation of motion are due to multiple insertions of vertex operators. When these insertion points approach each other, singularities follow from the operator product expansion of vertex operators, and they become sources of conformal symmetry breaking [10-13]. Thus the conformal invariance of the nonlinear sigma model on a two-sphere gives the equations of motion at the string tree level.

The next step is to incorporate effects of string loops which are represented by higher genus Riemann surfaces. The quantum effects of string loops may well be essential to determine string vacuum configurations. If one assumes conformal invariance of the nonlinear sigma model on a higher genus surface with a <u>fixed</u> metric, one obtains the equations for the background fields identical to those at the string tree level [14, 15]. On the other hand, the string loop corrections clearly exist in S-matrix elements and the loop-corrected equations of motion for string have been explicitly obtained in the S-matrix approach [16-18]. In the nonlinear sigma model, Fischler and Susskind have proposed a mechanism to obtain string loop corrections for equations of motion [19]. They suggested that divergences in the integration of modular parameters of Riemann surfaces are responsible for the conformal symmetry breaking. The suggestion has been studied with more explicit realizations recently [20-25]. Their suggestion can be interpreted as follows. Let us consider a h-loop tadpole amplitude $T_{(h)}$ of a string state in the background. In this paper, we shall employ the Schottky parametrization [26-29] of h-loop amplitudes, in which $V(SL(2, C))^{-1}$ can be naturally factored out of $T_{(h)}$

$$T_{(h)} = V(SL(2, C))^{-1} t_{(h)}. \tag{1.3}$$

Since V(SL(2, C)) is divergent, the loop corrected tadpole amplitude $\sum_{h=0}^{\infty} T_{(h)}$ vanishes provided $\sum_{h=0}^{\infty} t_{(h)}$ is not divergent. This requirement is the generalized conformal invariance where the conformal symmetry is maintained by cancelling divergences from different h (order of loops).

The purpose of our paper is to develop a systematic procedure for obtaining the loop-corrected string equations of motion from the nonlinear sigma model on higher genus Riemann surfaces. We shall provide a formalism to treat on the same footing the two origins of conformal symmetry breaking: coincidence singularities of several vertex operators and divergences on the boundary of moduli space of Riemann surfaces [24]. An operator formalism on Riemann surfaces with an arbitrary genus has been developed recently [30-33]. It enables us to express correlation functions on higher genus Riemann surfaces by free boson operators acting on Fock space. We shall show that a vacuum state vector $|\phi\rangle_{(h)}$ with the background ϕ for the genus h surface can be factored out of the scattering amplitudes under certain conditions. On the two-sphere, this factorization follows from a usual analysis in the interaction picture. On the higher genus surfaces, we use the above mentioned operator formalism and define "handle operators" which look like a kind of interaction term representing the effect of the addition of handles to the two-sphere. The breaking of conformal invariance are cancelled among various world sheet topologies by bringing them to the two-sphere with handle operators and vertex operators inserted.

The conformal invariance is broken when the nonlinear sigma model diverges at short distances on the two-dimensional world sheet. In collecting the short-distance divergences, we find it very useful to employ the Schottky parametrization [26-29], which expresses the genus h surface as a two-sphere with h-pairs of holes cut out and pairwise identified. In addition to the "positions" of h-pairs of holes, we have, as moduli, h complex parameters q_s $(s=1, \cdots, h)$ corresponding to the length and twist angle of the s-th handle. The conformal symmetry breaking determining the equations of motion is found to arise from short-distance divergences of handle positions. We argue that divergences of the q_s integration do not contribute to the determination of the equations of motion and should be regarded as a pathology of the string model. The Schottky parametrization separates moduli parameters naturally into the above two classes which play different roles in determining the equations of motion.

Divergences on boundaries of moduli space may be regarded as arising from operator product expansions among handle operators and vertex operators. Thus small handles with size comparable to the cut-off on the world sheet provide a new source of the conformal symmetry breaking. The fusion of a single vertex operator with handle operators gives a self-energy diagram as the coefficient of the short-distance divergence. The two-loop case is shown to contain a genuine two-loop self-energy in addition to the repetition of one-loop self-energies. We find that the one-particle pole term is properly subtracted and thus the two-loop self-energy is already one-particle irreducible. Similarly one can analyze the multiloop case by examining the integration regions of the Schottky parametrization in detail. By summing up the leading divergences of multiloop case, we can explicitly show that the one-loop self-energy is exponentiated to all orders.

We obtain the β-function by solving the renormalization group equation [34-38]. The fixed point condition (β=0) is found to give, to the four-point interaction level in the weak field expansion, the correct equations of motion which agrees with the result of the S-matrix approach [39] in the on-shell limit. A straight-forward computation of the next order reveals that the one-particle reducible parts are not correctly subtracted. We find that a nonlinear modification of the renormalization group equation provides a correct subtraction of one-particle reducible parts.

In sect. 2, the nonlinear sigma model on the two-sphere is analyzed to obtain the equations of motion at the string tree level. In sect. 3 the nonlinear sigma model on higher genus Riemann surfaces are treated by means of the operator formalism and the Schottky parametrization.

2. Nonlinear sigma model on a two-sphere

2.1 Short-distance divergence and OPE

A bosonic string propagating in curved spacetime with metric $G_{\mu\nu}$ is given by a nonlinear sigma model with an action*)

$$S = \frac{-1}{2\pi} \int d^2z \; \partial X^{\mu}(\bar{z})\bar{\partial}X^{\nu}(\bar{z}) \; G_{\mu\nu}(X), \tag{2.1}$$

where we use the conformal gauge and complex coordinate for the two-dimensional metric on the world sheet

$$ds^2 = 2g_{z\bar{z}} \; dzd\bar{z}, \qquad \sqrt{g} = 2g_{z\bar{z}}. \tag{2.2}$$

We can define the graviton field $H_{\mu\nu}$ on the flat metric $\eta_{\mu\nu}$

$$G_{\mu\nu}(X) = \eta_{\mu\nu} + 2\kappa \; H_{\mu\nu}(X) \tag{2.3}$$

with the gravitational coupling constant κ. Correspondingly we obtain a free action S_0 and an interaction part S_{int}

$$S = S_0 + S_{int}$$

$$S_0 = \frac{-1}{2\pi} \int d^2z \; \partial X^{\mu} \; \bar{\partial}X^{\nu} \; \eta_{\mu\nu} \tag{2.4}$$

*) Our convention for the string tension is $T=(2\pi\alpha')^{-1}\equiv(4\pi)^{-1}$.

$$S_{int} = -\frac{\kappa}{\pi}\int d^2z\, \partial X^\mu\, \bar\partial X^\nu\, H_{\mu\nu}(X)$$
$$= -\int dk\, h_{\mu\nu}(k) \int d^2z\, \partial X^\mu\, \bar\partial X^\nu\, e^{ikX}, \qquad (2.5)$$

where we introduced a Fourier transform and short-hand notations

$$h_{\mu\nu}(k) \equiv \frac{\kappa}{\pi}\int d^DX\, e^{-ikX}\, H_{\mu\nu}(X)$$
$$dk \equiv d^Dk/(2\pi)^D. \qquad (2.6)$$

The interaction part of the action is recognized as the graviton vertex operator $V(k, z) = \partial X^\mu \bar\partial X^\nu e^{ikX(z)}$ multiplied by the graviton wave function $h_{\mu\nu}(k)$. The nonlinear sigma model can be generalized to the string propagation on various backgrounds corresponding to vertex operators. The background of massless modes such as gravitons is most interesting for practical purposes and the separation of massless modes can be unambiguously done by expanding in powers of the slope parameter α'[40, 41]. In order to examine the mechanism of the short-distance divergence in the simplest context, however, we shall at first mainly take tachyon background in the closed bosonic string as an example

$$S_{int} = \int dk\, \phi(k) \int \frac{d^2w}{a^2} \sqrt{g}\, e^{ikX(w)}, \qquad (2.7)$$

where we introduced a two-dimensional short-distance cut-off parameter a [10] to make ϕ dimensionless (in two-dimensions) and we used a short-hand notation for the tachyon wave function $\phi(k) = \kappa\int d^DX\, e^{-ikX}\, \Phi(X)/\pi$. We can easily generalize the model to include backgrounds of massless and/or massive modes.

In this section we examine the nonlinear sigma model on S^2, namely of the string tree level. In order to define the vertex operator precisely, we introduce the general coordinate invariant (in two-dimensions) regularization by using a cut-off a in geodesic distances*)

---

*) To discuss terms with derivatives of $g^{z\bar z}$ which are relevant for gauge terms in massless fields, it is better to use more elaborate regularization such as in ref. [42].

$$<0|\hat{X}(z)\hat{X}(w)|0> = -\log(|z-w|^2+\varepsilon^2)$$

$$\varepsilon^2 = a^2 \cdot \tfrac{1}{2}g^{z\bar{z}} , \qquad (2.8)$$

where the hats denote operators acting on the Fock space of string modes defined on the complex plane

$$\hat{X}(z) = \hat{q} - i\hat{p}\,\ell n|z|^2 + i\sum_{n\neq 0}\frac{1}{n}(\hat{\alpha}_n z^{-n} + \hat{\bar{\alpha}}_n \bar{z}^{-n}) \qquad (2.9)$$

$$\hat{\alpha}_n|0> = 0, \qquad \hat{\bar{\alpha}}_n|0> = 0 \qquad n>0, \qquad \hat{p}\,|0> = 0.$$

Therefore we find the action with the tachyon background in terms of the normal ordered product: $:e^{ik\hat{X}}:$

$$S_{int} = \int dk\phi(k)\int d^2w\left(\frac{a^2}{2}g^{w\bar{w}}\right)^{\frac{k^2}{2}-1} :e^{ik\hat{X}(w)}: \quad . \qquad (2.10)$$

We see that the tachyon vertex operator $:e^{ik\hat{X}(w)}:$ acquires the anomalous dimension $\Delta = k^2/2$. In general we shall consider various vertex operators $V_i(z)$ and denote its anomalous dimension Δ_i.

Let us consider the n-point scattering amplitude in the background ϕ

$$A(1,\cdots,n) = 2\pi\int\frac{\prod_{i=1}^{n} d^2z_i\varepsilon^{2\Delta_i-2}}{V(SL(2,\ C))}\left(\frac{\kappa}{\pi}\right)^{n-2}<0|\prod_{i=1}^{n}\hat{V}_i(z)\ e^{\hat{S}_{int}}|0> \qquad (2.11)$$

$$V(SL(2,\ C)) = d^2z_a d^2z_b d^2z_c |z_a-z_b|^{-2}|z_b-z_c|^{-2}|z_c-z_a|^{-2}. \qquad (2.12)$$

To evaluate $<0|\prod_{i=1}^{n}\hat{V}_i(z_i)e^{\hat{S}_{int}}|0>$, short-distance singularities among V_i's including those from $e^{\hat{S}_{int}}$ must be regularized. We shall assume that S_{int} is adiabatically cut-off near $z \to \infty$. Then the integrand of $z_1, \cdots, z_n$ for $|z_n| >> 1$ factorizes as

$$<0|\prod_{i=1}^{n}\hat{V}_i(z_i)\ e^{\hat{S}_{int}}|0> \xrightarrow[|z_1|,\cdots,|z_n|>>1]{} <0|\prod_{i=1}^{n}\hat{V}_i(z_i)|\phi>$$

$$|\phi\rangle = \sum_{i=1}^{\infty}\frac{1}{N!}\int\prod_{j=1}^{N} dk_i\phi(k_i)\int\prod_{j=1}^{N} d^2w_j\left(\frac{a^2}{2}g^{w_j\bar{w}_j}\right)^{\frac{k_j^2}{2}-1} :e^{ik_jX(w_j)}:|0\rangle \,, \tag{2.13}$$

In the following we are interested only in the property of $|\phi\rangle$ which can be probed by inserting a finite number of vertex operators near $z \to \infty$. These insertions are most conveniently treated in terms of the so-called τ-functions [43, 30]

$$\tau(x) = \langle x, \bar{x}|\phi\rangle , \tag{2.14}$$

where the coherent state $\langle x,\bar{x}|$ is defined as an eigenstate of annihilation operators with eigenvalue x_n, $\bar{x}_n$

$$\begin{aligned}\langle x,\bar{x}|\hat{X}(z) &= \{\mathfrak{X}(z)+\log\frac{1}{|z|^2}\frac{\partial}{\partial x_0}+\sum_{n\geq 1}\frac{1}{n}(z^{-n}\frac{\partial}{\partial x_n}+\bar{z}^{-n}\frac{\partial}{\partial \bar{x}_n})\}\langle x,\bar{x}| \\ \mathfrak{X}(z) &= x_0 + \sum_{n\geq 1}(x_n z^n+\bar{x}_n\bar{z}^n).\end{aligned} \tag{2.15}$$

The vertex operators can be expressed as differential operators acting on the τ-function. The τ-function is also given by a path integral with a boundary condition specified at infinity

$$\begin{aligned}\tau(x) &= \int_{X\in\mathfrak{S}(x,\,\bar{x})}[dX]\; e^{S_0[X]+S_{int}[X]} \\ \mathfrak{S}(x,\bar{x}) &= \{X(z,\bar{z})\,|\,X(z,\bar{z}) \to \mathfrak{X}(z,\bar{z}) \text{ as } z \to \infty\}.\end{aligned} \tag{2.16}$$

We will compute the short-distance divergence of the nonlinear sigma model perturbatively in powers of κ (weak field expansion). The first term is already given by eq.(2.10). The short-distance divergence of the second term can be evaluated by means of the operator product expansion between vertex operators [11]

$$\begin{aligned}&:e^{ik_1\hat{X}(z_1)}::e^{ik_2\hat{X}(z_2)}: \\ &= (|z_1 - z_2|^2 + \varepsilon^2)^{k_1k_2}:e^{i(k_1+k_2)\hat{X}(z_1)}: + \cdots .\end{aligned} \tag{2.17}$$

By integrating over z_2, we find the short-distance divergence

$$\langle x,\bar{x}|\frac{1}{2}\int\frac{d^2z_1}{a^2}\sqrt{g}\; e^{ik_1X(z_1)}\int\frac{d^2z_2}{a^2}\sqrt{g}\; e^{ik_2X(z_2)}|0\rangle \tag{2.18}$$

$$= \frac{1}{2}\int d^2z(\frac{a^2}{2}g z\bar{z})^{\frac{1}{2}k^2-1}\frac{-2\pi}{2k_1k_2+2}\; e^{ik\mathfrak{X}(z)}\Big|_{k=k_1+k_2} + \cdots ,$$

where dots denote terms with functional form of $\mathfrak{X}(z)$ other than $e^{ik\mathfrak{X}(z)}$. Therefore we obtain to this order

$$\tau(x) = 1 + \int dk\int d^2z(\frac{a^2}{2}g z\bar{z})^{\frac{k^2}{2}-1}e^{ik\mathfrak{X}(z)}W(\phi,k) + \cdots \tag{2.19}$$

$$W(\phi,k) = \phi(k) + \frac{1}{2!}\int dk_1dk_2\phi(k_1)\phi(k_2)\delta(k-k_1-k_2)\frac{-2\pi}{2k_1k_2+2} + \cdots . \tag{2.20}$$

The short-distance divergence of the N-th order term ($N\geq3$) can be evaluated using rescaled variables z'

$$z_i' = (z_i - z_1)/(z_2 - z_1), \qquad z'_1 = 0, \qquad z'_2 = 1. \tag{2.21}$$

The integration over z_2 with the fixed rescaled variables z'_i gives

$$\langle x,\bar{x}|\frac{1}{N!}\int\prod_{i=1}^{N}\frac{d^2z_i}{a^2}\sqrt{g}\; e^{ik_i\hat{X}(z_i)}|0\rangle$$

$$= \frac{1}{N!}\int d^2z_1\; \varepsilon^{\sum_{i=1}^{N}(k_i^2-2)}e^{ik\mathfrak{X}(z_1)} \tag{2.22}$$

$$\times\int d^2z_2(|z_2-z_1|^2+\varepsilon^2)^{\sum_{1\leq i<j\leq N}k_ik_j+N-2}\; t_0^{(N+1)}(-k,k_1,\cdots,k_N)+\cdots$$

$$= \frac{1}{N!}\int d^2z_1\varepsilon^{k^2-2}e^{ik\mathfrak{X}(z_1)}\frac{-2\pi}{2\sum_{1\leq i<j\leq N}k_ik_j+2N-2}\cdot t_0^{(N+1)}(-k,k_1,\cdots,k_N)+\cdots ,$$

where $t_0^{(N+1)}$ is the dual N+1 point amplitude

$$t_0^{(N+1)}(-k,k_1,\cdots,k_N) = \int\prod_{i=3}^{N}d^2z_i'\prod_{1\leq j<\ell\leq N}|z_j'-z_\ell'|^{2k_jk_\ell} \tag{2.23}$$

with $z_1' = 0$ and $z_2' = 1$.

In particular for N=3, we obtain the dual four point amplitude

$$t_0^{(4)}(-k,\; k_1,\; k_2,\; k_3) = \pi\frac{\Gamma(-1-k_1k_2-k_2k_3)\Gamma(1+k_1k_2)\Gamma(1+k_2k_3)}{\Gamma(2+k_1k_2+k_2k_3)\Gamma(-k_1k_2)\Gamma(-k_2k_3)}. \tag{2.24}$$

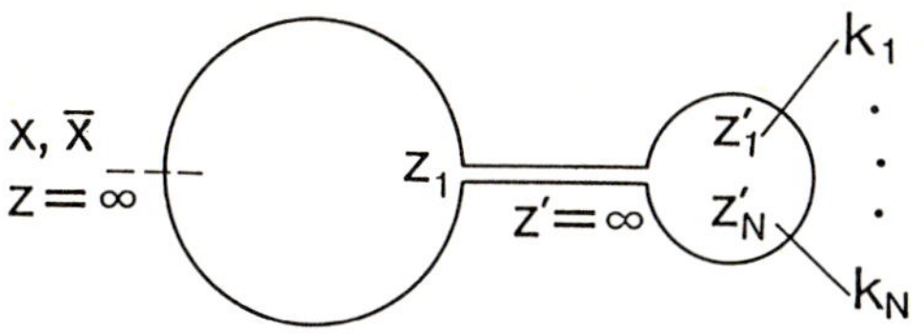

Fig. 2.1 The short-distance divergence of the N-th order term in the weak field expansion of the τ-function. The coherent state $<x, \bar{x}|$ is placed at $z=\infty$ and the conformal symmetry breaking appears as the conformal factor $g^{z\bar{z}}$ dependence at z_1.

The short-distance divergence of the N-th order term is a manifestation of the conformal symmetry breaking arising from vertex operator positions being close together. This mechanism can be visualized by a pinching of a topologically trivial cycle around $z=\infty$ as illustrated in Fig. 2.1. Therefore we find to all orders of the weak field expansion that terms proportional to $e^{ik\mathfrak{X}(z)}$ have powers of $a^2 g^{z\bar{z}}/2$ corresponding to the anomalous dimension of the tachyon vertex operator as in eq.(2.19) and that the W-function is given by

$$W(\phi,k) = \sum_{N=1}^{\infty} \frac{1}{N!} \int \prod_{i=1}^{N} dk_i \phi(k_i) \delta(k-\Sigma k_i) \frac{-2\pi}{2\sum_{1\leq i<j\leq N} k_i k_j + 2N-2} \cdot t_0^{(N+1)}(-k, k_1, \cdots, k_N) \quad . \tag{2.25}$$

2.2 Renormalization group equation and β-function

From the point of view of two-dimensional field theory, the state vector $|\phi>$ or equivalently its coherent state representation $\tau(x) = <x, \bar{x}|\phi>$ is the generating function for correlation functions in two-dimensions, and the tachyon wave function $\phi(k)$ serves as a coupling constant for each k. In the above we have computed the cut-off dependence of $\tau(x)$ and the results are summarized in eqs.(2.19) and (2.25). In order to take a limit $a \to 0$ in the non-linear sigma model in two-dimensions, its coupling constants $\phi(k)$ must be varied under a change of the cut-off scale a so that the explicit a - dependence of $\tau(x)$ in eq.(2.19) is compensated. This requirement gives the following first-order differential equation for $\phi(k)$.

$$\left(\frac{k^2}{2} - 1\right) W(\phi,k) + \int dk'\beta(k')\frac{\delta}{\delta\phi(k')} W(\phi,k) = 0,$$

$$\frac{1}{2} a\frac{d}{da} \phi(k) = \beta(k), \tag{2.26}$$

$$|\phi\rangle = |0\rangle + \int dk W(\phi,k)\int d^2z \, \varepsilon^{k^2-2} :e^{ik\hat{X}(z)}: |0\rangle + \cdots .$$

When the non-linear sigma model is conformally invariant, its coupling constant $\phi(k)$ must be independent of a. Thus $\beta(k) = 0$ is the condition for the tachyon background to preserve the conformal invariance on S^2. In this section, we are going to solve eq.(2.26) to obtain $\beta(k)$ as a functional of $\phi(k)$.

Let us try to solve eq.(2.26) iteratively by powers of the wave function $\phi(k)$. The power series expansion of $W(\phi,k)$ is given in eq.(2.25). Expanding $\beta(k)$ in powers of $\phi(k)$,

$$\beta(k) = \sum_{N=1}^{\infty}\frac{1}{N!}\int \prod_{i=1}^{N} dk_i\phi(k_i)\delta(k-\sum_{j=1}^{N} k_j)\beta^{(N+1)}(-k,k_1,\cdots,k_N), \tag{2.27}$$

and substituting it into eq.(2.26), we obtain recurrence relations for the expansion coefficients $\beta^{(N+1)}$ as

$$\beta^{(2)}(-k,k) = -\frac{k^2}{2} + 1 \tag{2.28}$$

$$\beta^{(3)}(-k,k_1,k_2) = \frac{2\pi}{2k_1k_2+2}\left(\frac{k^2}{2} - 1 + \beta^{(2)}(-k_1,k_1) + \beta^{(2)}(-k_2,k_2)\right)$$
$$\times\, t_0^{(3)}(-k,k_1,k_2) \qquad (k=k_1+k_2) \tag{2.29}$$

$$\beta^{(4)}(-k,k_1,k_2,k_3)$$
$$= \frac{2\pi}{2k_1k_2+2k_2k_3+2k_3k_1+4}\left[\frac{k^2}{2}-1+\beta^{(2)}(-k_1,k_1)+\beta^{(2)}(-k_2,k_2)+\beta^{(2)}(-k_3,k_3)\right]$$
$$\times\, t_0^{(4)}(-k,k_1,k_2,k_3)$$
$$+ \sum_{j=1}^{3}\frac{2\pi}{2k_j(k_{j+1}+k_{j+2})+2}\beta^{(3)}(-k_{j+1}-k_{j+2},k_{j+1},k_{j+2})$$
$$\times\, t_0^{(3)}(-k,k_j,k_{j+1}+k_{j+2})$$
$$k=k_1+k_2+k_3$$
$$k_{j+3} = k_j \tag{2.30}$$

Comparing eqs.(2.20) and (2.25), one can see $t_0^{(3)}(-k,k_1,k_2) = 1$ provided $k = k_1 + k_2$. By substituting eq.(2.28) into eq.(2.29) we obtain $\beta^{(3)}(-k,k_1,k_2) = \pi$ and the β-function to the order of ϕ^2 is given by

$$\beta(k) = (-\frac{k^2}{2} + 1)\phi(k) + \frac{\pi}{2!}\int dk_1 dk_2 \phi(k_1)\phi(k_2)\delta(k-k_1-k_2) + O(\phi^3). \tag{2.31}$$

Note that, although the quadratic term in $W(\phi,k)$ has a prefactor $\frac{-2\pi}{2k_1k_2+2}$ which is non-local in the target spacetime, this non-locality has disappeared in the quadratic term in $\beta(k)$. As one can see in eq.(2.29), the cancellation of the momentum dependence of this prefactor is due to $\beta^{(2)}$'s in the parenthesis of the right-hand side. From the viewpoint of two-dimensional field theory, $\beta^{(2)}$'s in the right-hand side of eq.(2.29) serve to subtract terms due to divergences in sub-diagrams.

Let us proceed to the next order of the iteration. Substituting eq.(2.30), we obtain

$$\beta^{(4)}(-k,k_1,k_2,k_3) = \pi\left(t_0^{(4)}(-k,k_1,k_2,k_3) + \frac{2\pi}{2k_1(k_2+k_3)+2} + \frac{2\pi}{2k_2(k_3+k_1)+2} + \frac{2\pi}{2k_3(k_1+k_2)+2}\right). \tag{2.32}$$

Especially when the four external states are put on the mass shell ($k^2=k_1{}^2=k_2{}^2=k_3{}^2=2$), the above becomes

$$\beta^{(4)}(-k,k_1,k_2,k_3) = \pi\left(t_0^{(4)}(-k,k_1,k_2,k_3) - \frac{2\pi}{(k_2+k_3)^2-2} - \frac{2\pi}{(k_3+k_1)^2-2} - \frac{2\pi}{(k_1+k_2)^2-2}\right). \tag{2.33}$$

In the on-shell limit, $t_0^{(4)}$ becomes the dual four point amplitude as given in eqs. (2.23) and (2.24), and its one-tachyon pole terms are correctly subtracted out in eq.(2.33) by the last three terms in the parenthesis of the right-hand side. The resulting $\beta^{(4)}$ agrees with the effective four point coupling of tachyons computed in the S-matrix approach [39]. In the terminology of two-dimensional field theory, one-tachyon pole terms correspond to divergences in sub-diagrams and they are eliminated by solving the renormalization group equation (2.26).

As we have seen so far, to the cubic order of the power series expansion, the β-function obtained by solving the renormalization group equation correctly reproduces the effective interactions of tachyons computed in the S-matrix approach at least in the mass shell limit. Now let us examine whether this feature persists to an arbitrary order of the power series expansion.

In the field theory of point particles, there is a well-known prescription to recover one-particle irreducible interactions from scattering amplitudes which may contain one-particle pole terms; the Legendre transformation between a generating function of scattering amplitudes and an effective action. We will compare this procedure with the renormalization group equation to understand the relation between the S-matrix approach and the computation of the β-function.

Let us rewrite the renormalization group equation (2.26) so that we can compare it with the Legendre transformation. Separating the linear term $(-\frac{k^2}{2}+1)\phi(k)$ in $\beta(k)$ as

$$\beta(k) = (-\frac{k^2}{2}+1)\phi(k) + \tilde{\beta}(k), \tag{2.34}$$

the renormalization group equation (2.26) takes the form

$$\tilde{\beta}(k) = \tilde{W}(\phi, k) - \int dk' \frac{\delta\tilde{W}(\phi,k)}{\delta\phi(k')} \cdot \frac{1}{k'^2/2-1}\tilde{\beta}(k'), \tag{2.35}$$

$$W(\phi,k) = \sum_{N\geq 2}\frac{1}{N!}\int\prod_{i=1}^{N} dk_i\ \phi(k_i)\prod t_0^{(N+1)}(-k,k_1,\cdots,k_N)\delta(k-\sum_{i=1}^{N} k_i), \tag{2.36}$$

when the support of $\phi(k)$ is limited near the mass-shell $k^2\simeq 2$.

In the S-matrix approach, $\tilde{W}(\phi, k)$ in the above may be related to a generating function $\mathcal{W}(J)$ for tachyon scattering amplitudes as

$$\tilde{W}(\phi,\ k) = (-\frac{k^2}{2}+1)\frac{\delta \mathcal{W}}{\delta J(-k)}(J(k')=(-\frac{k'^2}{2}+1)\phi(k')) + (-\frac{k^2}{2}+1)\phi(k). \tag{2.37}$$

An effective action $\Gamma(\phi)$ is given by the Legendre transform of $\mathcal{W}(J)$. By separating the linear term $(-k^2/2+1)\phi(k)$ in $\delta\Gamma/\delta\phi(k)$ as

$$\frac{\delta\Gamma}{\delta\phi(-k)} = (-\frac{k^2}{2}+1)\phi(k) + \tilde{j}(k), \tag{2.38}$$

we obtain the following functional equation for the interaction term $\tilde{j}(k)$

$$\tilde{j}(k) = \tilde{W}(\phi(k') - \frac{1}{k'^2/2-1}\tilde{j}(k'),\ k). \tag{2.40}$$

The most significant difference between eqs.(2.35) and (2.40) is that the former is at most linear in $\tilde{\beta}(k)$ while the latter is a nonlinear equation for $\tilde{j}(k)$. Thus it is suspected that $\tilde{\beta}(k)$ and $\tilde{j}(\phi,k)$ given by solving these equations will disagree in higher orders of the weak field expansion. In fact we have found by explicit calculations that, although $\tilde{\beta}(k)$ and $\tilde{j}(\phi,k)$ agree to the cubic order of ϕ, the quartic terms (corresponding to five point interactions of the tachyon) are different. In $\tilde{j}(\phi,k)$, one-tachyon pole terms in the five point amplitude $t_0^{(5)}(-k,k_1,k_2,k_3,k_4)$ is correctly subtracted out while they are over-subtracted in $\tilde{\beta}(k)$.

Therefore it seems that, in higher order terms of the ϕ-expansion, the β-function $\beta(k)$ given by solving the renormalization group equation (2.26) does not reproduce the equation of motion for the tachyon by $\beta(k)=0$. One possible way out of this problem will be the following. So far we have considered divergences coming from the region where insertion points for external vertex operators are close to each other. However in higher orders in the weak field expansion, we should take into account other types of divergences, e.g. those occuring when the background tachyon vertex operators approach to the external vertex operators. These various divergences should be properly renormalized order by order in the weak field expansion.

This procedure will give rise to nonlinear modifications to the renormalization group equation (2.35). Although we have not explored this possibility in detail, it is expected that such nonlinear modifications to the renormalization group equation will resolve the discrepancy between the tachyon equation of motion and $\beta(k)=0$ in higher order terms.

In the following we shall concentrate lower order terms in the ϕ-expansion so that the above issue can be neglected.

3. NL-σ model on Riemann surfaces with handles

3.1 One-loop case (torus)

The n-point one-loop (torus topology) amplitude in the background ϕ is given by

$$A_1(1, \cdots, n) = \int_F \frac{d^2\tau}{(\mathrm{Im}\tau)^2} Z(\tau,\bar{\tau})\left(\frac{\kappa}{\pi}\right)^n \int \prod_{i=1}^{n} d^2u_i \times \left\langle \prod_{i=1}^{n} \varepsilon^{2\Delta_i - 2} V_i(u_i) \exp S_{int} \right\rangle_\tau , \tag{3.1}$$

where the subscript 1 denotes one-loop (genus one) and

$$Z(\tau,\bar{\tau}) = \frac{1}{2}(8\pi^2)^{-13}(\sqrt{\mathrm{Im}\tau}\,|\eta(\tau)|^2)^{-24}$$

$$\eta(\tau) = e^{\frac{i\pi\tau}{12}} \prod_{n=1}^{\infty} (1-e^{2\pi i n\tau}).$$

The correlation function $\langle V(u_1)\cdots\rangle_\tau$ is evaluated on the torus which is obtained by identifying opposite sides of the parallelogram with sides defined by 1 and τ.

First we seek for a change of variables which maps the parallelogram coordinate u to a coordinate z on a two-sphere by requiring

(1) Translational invariance $u_i \to u_i + \varepsilon$ for $i=1 \sim n$ becomes an element of SL(2,C) transformation in the z-coordinate.

(2) $z(u)$ fixes three of $z_1, \cdots, z_n$ e.g. z_1, z_2, and z_3. We find the following conformal mapping as the solution

$$\frac{z(u)-z_1}{z(u)-z_3}\frac{z_2-z_3}{z_2-z_1} = \frac{e^{2\pi iu}-e^{2\pi iu_1}}{e^{2\pi iu}-e^{2\pi iu_3}}\frac{e^{2\pi iu_2}-e^{2\pi iu_3}}{e^{2\pi iu_2}-e^{2\pi iu_1}} , \tag{3.2}$$

where $z_1=z(u_1)$ etc. Then we find

$$d^2u_2\ d^2u_3 = (2\pi)^4 \left|\frac{dz_1}{du_1}\right|^{-2}\left|\frac{dz_2}{du_2}\right|^{-2}\left|\frac{dz_3}{du_3}\right|^{-2}\frac{1}{|z_0-z_\infty|^4}$$
$$\prod_{1\le i<j\le 3} |z_i-z_j|^2\ d^2z_0 d^2z_\infty ,$$

where z_0 and z_∞ correspond to $u=-i\infty$ and $+i\infty$ and are defined by

$$\frac{z_\infty-z_1}{z_\infty-z_3}\cdot\frac{z_2-z_3}{z_2-z_1} = \frac{e^{2\pi iu_2}-e^{2\pi iu_3}}{e^{2\pi iu_2}-e^{2\pi iu_1}}$$
$$\frac{z_0-z_1}{z_0-z_3}\cdot\frac{z_\infty-z_3}{z_\infty-z_1} = e^{-2\pi i(u_3-u_1)} . \tag{3.3}$$

Geometrically z_0 and z_∞ are two limiting points inside two circular holes excluded from the integration region of z variables and correspond to the position of attached handles on the sphere as shown in Fig. 3.1. As described in ref. [47], the torus can be coordinatized as a sphere with two holes cut out and identified (Schottky parametrization)[26-29]. The above conformal mapping connects the Schottky parametrization z with the standard parallelogram coordinate u.

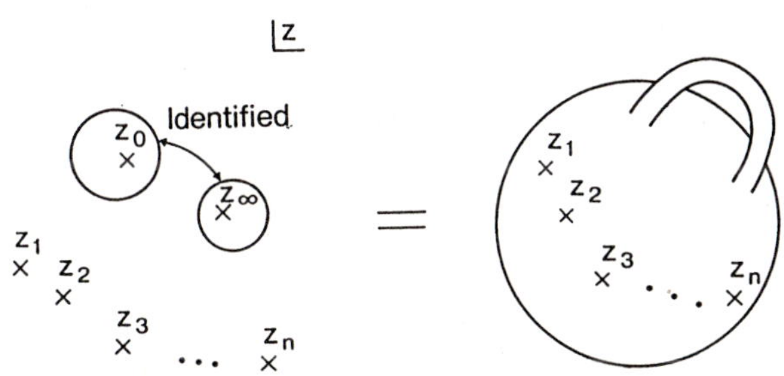

Fig. 3.1 The Schottky parametrization for a torus. Identifying two holes on a sphere gives a torus.

Since we are using a covariant regularization such as in eq.(2.8), $\varepsilon^{2\Delta-2}V(u)$ is a scalar under two dimensional reparametrizations. We thus find

$$\prod_{i=2}^{n} d^2u_i < \prod_{i=1}^{n} \varepsilon^{2\Delta_i-2} V_i(u_i)\, e^{S_{int}} >_\tau$$
$$= \frac{\prod_{i=1}^{n} d^2z_i \varepsilon^{2\Delta_i-2}}{V(SL(2,C))} \frac{d^2z_0 d^2z_\infty}{|z_0-z_\infty|^4} < \prod_{i=1}^{n} V_i(z_i)\, e^{S_{int}} >_\tau . \tag{3.4}$$

We notice that the SL(2,C) gauge volume is factored out

$$V(SL(2,C)) = \frac{d^2z_1 d^2z_2 d^2z_3}{|z_1-z_2|^2 |z_2-z_3|^2 |z_3-z_1|^2}. \tag{3.5}$$

We can fix any three points out of $z_1, \cdots, z_n$ instead of z_1, z_2, z_3. With this Schottky parametrization, we can naturally compare conformal symmetry breaking of the nonlinear sigma-model on topologically different surfaces, since we can compare the dependence on the conformal factor at the same coordinate point z of the two-sphere which represents surfaces with various genera.

Up to now we have merely changed a coordinate system on the torus. The next step is to construct a handle operator e^G which enables us to convert the correlation function on the torus to an operator expression defined on the two-sphere

$$< \prod_{i=1}^{n} V_i(z_i)\, e^{S_{int}} >_\tau = <0| \prod_{i=1}^{n} \hat{V}_i(z_i)\, e^{\hat{S}_{int}}\, e^G |0>, \tag{3.6}$$

where hats denote operators defined on the two-sphere. Using the operator formulation on Riemann surfaces [30-33], we describe in Appendix B a method to obtain e^G for Riemann surfaces with arbitrary genera. We find explicitly for the torus

$$G = [\frac{-1}{8\pi \mathrm{Im}\tau}(\hat{X}(z_0)-\hat{X}(z_\infty))^2 - \{ \sum_{n=1}^{\infty} \frac{nq^n}{1-q^n} \frac{1}{n!} \frac{\partial^{n-1}}{\partial z_0^{n-1}}((z_0-z_\infty)^n \frac{\partial \hat{X}(z_0)}{\partial z_0})$$

$$\times \frac{1}{n!} \frac{\partial^{n-1}}{\partial z_{\infty}^{n-1}}((z_{\infty}-z_0)^n \frac{\partial \hat{X}(z_{\infty})}{\partial z_{\infty}}) + c.\ c.\}]_{creation} \tag{3.7}$$

$$q = e^{2\pi i \tau}$$

where the suffix "creation" means that only the creation operators ($\alpha_n, \bar{\alpha}_n, n<0$) are retained$^{*)}$, since the annihilation operators are irrelevant when acting on the Fock vacuum.

Although we have succeeded in constructing an operator formulation of the n-point amplitude in the Schottky parametrization for the torus, we need to point out that the integration over z_0 and z_{∞} is restricted to a region, whose extent depends on $z_1, \cdots, z_n$ and τ. The precise restriction is given in ref. [47] . For our purpose of discussing the nonlinear sigma-model on the torus, we are interested in the situation where the vertex operator insertions at $u_1, \cdots, u_n$ are close together. In that limit, we can choose a coordinate in the Schottky parametrization such that all the corresponding vertex positions $z_1, \cdots, z_n$ are near infinity: $|z_1|, \cdots, |z_n| >> 1$, $z_0, z_{\infty} \sim 1$. Since the vertex positions are far away from the handle positions z_0 and z_{∞}, we find that the integration of z_0 and z_{∞} receives no restriction in this limit as described more precisely in ref. [47] . Thus we find that the n-point amplitude on the torus can be rewritten in a form similar to the tree amplitude

$$A_1(1,\cdots,n) = 2\pi \int \frac{\prod_{i=1}^{n} d^2 z_i}{V(SL(2,C))} (\frac{\kappa}{\pi})^{n-2} \langle 0| \prod_{i=1}^{n} \varepsilon^{2\Delta_i - 2} \hat{V}_i(z_i) |\phi\rangle_{(1)}, \tag{3.8}$$

where the "torus vacuum state" $|\phi\rangle_{(1)}$ with the background ϕ can be factored out

$$|\phi\rangle_{(1)} = 2\pi(\frac{\kappa}{\pi})^2 \int_F \frac{d^2\tau}{(\mathrm{Im}\tau)^2} Z(\tau,\bar{\tau}) \mathrm{Im}\tau \frac{d^2 z_0 d^2 z_{\infty}}{|z_0 - z_{\infty}|^4} e^{\hat{S}_{int}} e^G |0\rangle. \tag{3.9}$$

---

*) Instead, we can define the handle operator as the normal ordered one $:e^G:$, if the annihilation operator are retained

We will denote the state vector $|\phi\rangle$ with the background on the sphere in eq.(2.13) as $|\phi\rangle_{(0)}$ hereafter. Since we succeeded in factoring out the state vectors $|\phi\rangle_{(0)}$ and $|\phi\rangle_{(1)}$, we are naturally led to propose a principle of superposition to define a vacuum state vector $|\phi\rangle_{\infty}$ of a generalized nonlinear sigma model on sum over Riemann surfaces with various genera: $|\phi\rangle_{\infty}=|\phi\rangle_{(0)}+|\phi\rangle_{(1)}+\cdots$. This suggests that the conformal symmetry breaking of the nonlinear sigma model should be cancelled among surfaces with various genera.

Using the torus vacuum state with the background, we can define the τ-function for the nonlinear sigma model on the torus

$$\tau_{h=1}(x) = \langle x,\ \bar{x}|\phi\rangle_{(1)}, \tag{3.10}$$

where $\langle x,\ \bar{x}|$ is the coherent state defined in eq.(2.15). We shall evaluate the short-distance divergences perturbatively in powers of κ (weak field expansion)

$$\tau_{h=1}(x) = \sum_{N=0}^{\infty} \tau_{h=1}^{N}(x). \tag{3.11}$$

The zeroth order term comes from the torus vacuum without the background

$$\begin{aligned}\tau_{h=1}^{N=0}(x) &= \langle x,\ \bar{x}|\phi = 0\rangle_{(1)} \\ &= 2\pi\left(\frac{\kappa}{\pi}\right)^2\int_F \frac{d^2\tau}{(\mathrm{Im}\tau)^2}Z(\tau,\ \bar{\tau})\,\mathrm{Im}\tau\frac{d^2z_0d^2z_\infty}{|z_0-z_\infty|^4}\langle x,\ \bar{x}|\ e^G\ |0\rangle.\end{aligned} \tag{3.12}$$

We find that the handle operator can be expanded*) in powers of z_0-z_∞

$$e^G = 1 - \frac{|z_0-z_\infty|^2}{4\pi\mathrm{Im}\tau}\,[\partial\hat{X}(z_0)\bar{\partial}\hat{X}(z_0)]_{\mathrm{creation}} + \cdots. \tag{3.13}$$

---

*) The terms with unequal powers of z_0-z_∞ and $\bar{z}_0-\bar{z}_\infty$ disappears upon integration over z_0 and z_∞, and hence we omitted to write them explicitly.

The first and the second terms correspond to the tachyon and the dilaton vertex operators, respectively, at vanishing momentum. Only these two terms give rise to the short-distance divergences as $z_\infty \to z_0$. Since the limiting points z_0 and z_∞ represent the handle positions on the two-sphere in the Schottky parametrization, we introduce the same regularization as that for the short-distance singularity of vertex operators on the two-sphere in eq.(2.8)

$$|z_0-z_\infty|^{-4} \longrightarrow (|z_0-z_\infty|^2 + \varepsilon^2)^{-2} \tag{3.14}$$

Then we find the short distance divergences due to tachyon and dilaton tadpoles as illustrated in Fig. 3.2

$$<x,\bar{x}|\phi = 0>_{(1)} \simeq \int d^2z_0 \{\frac{1}{\varepsilon^2} 2\pi\kappa T_\phi + \ln\frac{1}{\varepsilon^2} \partial X \bar{\partial} X \frac{\kappa}{2\pi} T_D\}, \tag{3.15}$$

where tachyon and dilaton tadpole amplitudes are given by

$$\begin{aligned} T_\phi &= \frac{\kappa}{\pi} \int_F \frac{d^2\tau}{(\mathrm{Im}\tau)^2} Z(\tau,\bar{\tau}) \int d^2u <:e^{ikX(u)}:>_\tau \Big|_{k=0} \\ &= \frac{\kappa}{\pi} \int_F \frac{d^2\tau}{(\mathrm{Im}\tau)^2} Z(\tau,\bar{\tau}) \mathrm{Im}\tau, \\ T_D &= \frac{\kappa}{\pi} \int_F \frac{d^2\tau}{(\mathrm{Im}\tau)^2} Z(\tau,\bar{\tau}) \int d^2u <:(-\partial X\bar{\partial}Xe^{ikX(u)}):>_\tau \Big|_{k=0} \\ &= \kappa \int_F \frac{d^2\tau}{(\mathrm{Im}\tau)^2} Z(\tau,\bar{\tau}) = -\kappa\Lambda_1. \end{aligned}$$

We have denoted the cosmological constant at one-loop order as Λ_1.

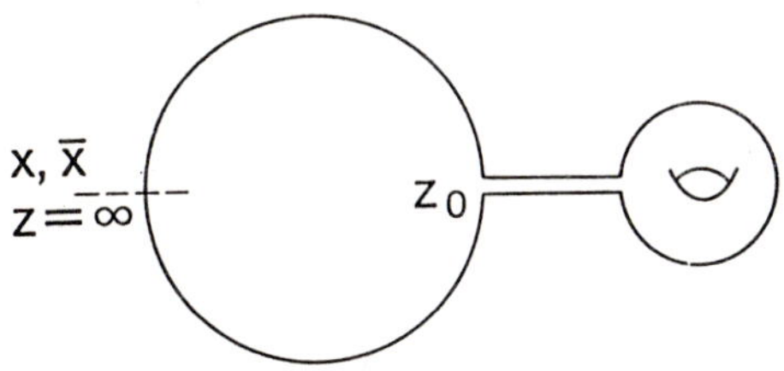

Fig. 3.2 Tachyon and dilaton tadpoles give short-distance divergence of the one-loop N=0 term. The coherent state $<x, \bar{x}|$ is placed at $z=\infty$ and the small handle at z_0 is responcible for the conformal symmetry breaking.

In order to find the short distance divergence due to the collision of vertex operators and handle operators, it is useful to expand the scalar Green's function in inverse powers of z, w, $\bar{z}$ and $\bar{w}$ as in ref. [47]

$$
\begin{aligned}
\langle X(z)X(w)\rangle &= -\log|z-w|^2 + \mathcal{G}(z,w;z_0,z_\infty) + \rho(z,\bar{z}) + \rho(w,\bar{w}) \\
\mathcal{G}(z,w;z_0,z_\infty) &= -2\sum_{n,m\geq 1}\frac{1}{nm}(Q_{nm}z^{-n}w^{-m} + \bar{Q}_{nm}\bar{z}^{-n}\bar{w}^{-m}) \\
&\quad + \sum_{n\geq 1}\frac{1}{n}\{Q_n(z^{-n}+w^{-n}) + \bar{Q}_n(\bar{z}^{-n}+\bar{w}^{-n})\} \\
&\quad - \frac{\pi}{2\mathrm{Im}\tau}[\sum_{n\geq 1}\frac{1}{n}\{A_n(z^{-n}-w^{-n}) - \bar{A}_n(\bar{z}^{-n}-\bar{w}^{-n})\}]^2,
\end{aligned}
\tag{3.16}
$$

where we allowed a function ρ of z, $\bar{z}$ (w, $\bar{w}$) alone which has no effect on various n-point functions if momentum conservation is used. The above expression can be used for arbitrary genus Riemann surfaces, and takes a particularly simple form in the case of the torus

$$
\begin{aligned}
&-2\sum_{n,m\geq 1}\frac{1}{nm}Q_{nm}z^{-n}w^{-m} + \sum_{n\geq 1}\frac{1}{n}Q_n(z^{-n}+w^{-n}) \\
&= \sum_{n\geq 1}\frac{1}{n}\frac{q^n}{1-q^n}\left\{\left(\frac{(z-z_0)(w-z_\infty)}{(z-z_\infty)(w-z_0)}\right)^n + \left(\frac{(z-z_\infty)(w-z_0)}{(z-z_0)(w-z_\infty)}\right)^n - 2\right\}, \\
&-\frac{\pi}{2\mathrm{Im}\tau}[\sum_{n\geq 1}\frac{1}{n}\{A_n(z^{-n}-w^{-n}) - \bar{A}_n(\bar{z}^{-n}-\bar{w}^{-n})\}]^2 \\
&= \frac{1}{8\pi\mathrm{Im}\tau}\{\ln\frac{(z-z_0)(w-z_\infty)}{(z-z_\infty)(w-z_0)} + \text{complex conjugate}\}^2.
\end{aligned}
\tag{3.17}
$$

We observe that the torus $\mathcal{G}$ function is SL(2, C) invariant, namely a function of the SL(2, C) invariant unharmonic ratio $(z-z_0)(w-z_\infty)\cdot(z-z_\infty)^{-1}(w-z_\infty)^{-1}$. Hence we can fix handle positions z_0 and z_∞ to arbitrary values, for instance with the rescaled variables z′ and w′ such as

$$
\begin{aligned}
z' &= (z-z_0)(z_\infty - z_0)^{-1} \\
\mathcal{G}(z, w; z_0, z_\infty) &= \mathcal{G}(z', w'; 0, 1)
\end{aligned}
\tag{3.18}
$$

We also note as $w' = z + \Delta z \to z$

$$
\mathcal{G}(z, z + \Delta z; z_0, z_\infty) \sim O(\Delta z)^2 \tag{3.19}
$$

The zeroth order term (3.12) of the τ-function in the weak field expansion can be expressed in terms of Q_{nm} and A_n as

$$<x,\bar{x}|\ e^G\ |0> = \exp[-\sum_{n,m\geq 1}(Q_{nm}x_nx_m+\bar{Q}_{nm}\bar{x}_n\bar{x}_m) + \frac{\pi}{2\mathrm{Im}\tau}\{\sum_{n\geq 1}(A_nx_n-\bar{A}_n\bar{x}_n)\}^2]\ . \tag{3.20}$$

The first order term (N=1) is then given by an insertion of the vertex operator

$$<x,\bar{x}|:e^{ik\hat{X}(z)}:e^G|0>$$
$$=e^{ik\mathcal{X}(z)}\exp\{ik\{\ell n\frac{1}{|z|^2}\frac{\partial}{\partial x_0}+\sum_{n=1}^{\infty}\frac{1}{n}(z^{-n}\frac{\partial}{\partial x_n}+\bar{z}^{-n}\frac{\partial}{\partial\bar{x}_n})\}]<x,\bar{x}|e^G|0>$$
$$=e^{ik\mathcal{X}(z)}<x_n=\frac{ik}{n}z^{-n},\ \bar{x}_n=\frac{ik}{n}\bar{z}^{-n}|e^G|0>+\cdots$$
$$=e^{ik\mathcal{X}(z)}\exp(k^2\mathcal{G}(z,\ \infty;\ z_0,\ z_\infty))+\cdots, \tag{3.21}$$

where dots denote terms with functional forms of $\mathcal{X}(z)$ other than the tachyon $\exp(ik\mathcal{X}(z))$. The function $\mathcal{G}$ in the exponent is given by

$$\mathcal{G}(z,\infty;z_0,z_\infty) = \sum_{n\geq 1}\frac{1}{n}(Q_nz^{-n}+\bar{Q}_n\bar{z}^{-n}) - \frac{\pi}{2\mathrm{Im}\tau}\{\sum_{n\geq 1}\frac{1}{n}(A_nz^{-n}-\bar{A}_n\bar{z}^{-n})\}^2$$

and more explicitly in the torus case by

$$\mathcal{G}(z,\infty;z_0,z_\infty) = \frac{1}{8\pi\mathrm{Im}\tau}\left(\ell n\left|\frac{z-z_0}{z-z_\infty}\right|^2\right)^2 + [\sum_{n\geq 1}\frac{1}{n}\frac{q^n}{1-q^n}\{(\frac{z-z_0}{z-z_\infty})^n+(\frac{z-z_\infty}{z-z_0})^n-2\}+\text{c.c.}] \equiv \mathcal{G}(z'), \tag{3.22}$$

where $\mathcal{G}$ is actually a function of the rescaled variable z' in eq.(3.18). This fact enables us to extract the short-distance divergence easily by a change of variables $(z_0,\ z_\infty,\ z) \to (z_0,\ z_\infty,\ z')$. The integral diverges logarithmically

$$\int\frac{d^2z_0d^2z_\infty}{|z_0-z_\infty|^4}\int d^2z\varepsilon^{k^2-2} \times <x,\bar{x}|:e^{ik\hat{X}(z)}:e^G|0>$$
$$= \int d^2z_0\varepsilon^{k^2-2}e^{ik\mathcal{X}(z_0)}\int\frac{d^2z_\infty}{|z_\infty-z_0|^2+\varepsilon^2}\int d^2z'\exp(k^2\mathcal{G}(z'))+\cdots .$$

Hence we obtain the first order term in eq.(3.11) as

$$\tau^{N=1}_{h=1}(x) = \int dk\phi(k)\int d^2z_0\varepsilon^{k^2-2}e^{ik\mathfrak{X}(z_0)}\frac{1}{2}\ln\frac{1}{\varepsilon^2}T_{h=1}(-k,k), \tag{3.23}$$

where $T_{h=1}(-k, k)$ denotes the one-loop two-point function

$$T_{h=1}(-k,k) = (2\pi)^2(\frac{\kappa}{\pi})\int_F\frac{d^2\tau}{(\mathrm{Im}\tau)^2}Z(\tau,\bar{\tau})\mathrm{Im}\tau\int d^2z'\, e^{k^2\mathcal{G}(z')}. \tag{3.24}$$

As shown in Fig. 3.3, this N=1 term represents a self-energy contribution.

The second order term (N=2) in the weak field expansion (3.11) is obtained from the insertion of two vertex operators

$$\begin{aligned}
&\langle x,\bar{x}|:e^{ik_1\hat{X}(z_1)}::e^{ik_2\hat{X}(z_2)}:\ e^G\ |0\rangle \\
&= e^{ik\mathfrak{X}(z_0)}|z_1-z_2|^{2k_1k_2}\langle x_n=\sum_{j=1}^{2}ik_j\frac{z_0^{-n}}{n}\ ,\bar{x}_n=\sum_{j=1}^{2}ik_j\frac{\bar{z}_0^{-n}}{n}\ |e^G|0\rangle+\cdots \\
&= e^{ik\mathfrak{X}(z_0)}|z_1-z_2|^{2k_1k_2}\exp\{-\sum_{1\le r<\ell\le 3}k_rk_\ell\mathcal{G}(z_r,z_\ell;z_0,z_\infty)\}+\cdots,
\end{aligned}$$

where $z_3=\infty$ and $k_3=-k=-k_1-k_2$. We need to extract the short-distance divergence of the integral

$$\begin{aligned}
I &= \int\frac{d^2z_\infty d^2z_1 d^2z_2}{(|z_0-z_\infty|^2+\varepsilon^2)^2}(|z_1-z_2|^2+\varepsilon^2)^{k_1k_2}e^{-\sum k_rk_\ell\mathcal{G}(z_r,z_\ell;z_0,z_\infty)} \\
&= \int d^2z_\infty(|z_0-z_\infty|^2+\varepsilon)^{k_1k_2}\int d^2z_1'd^2z_2'\left(|z_1'-z_2'|^2+\frac{\varepsilon^2}{|z_0-z_\infty|^2}\right)^{k_1k_2} \\
&\quad\cdot\exp(-\sum k_rk_\ell\mathcal{G}(z_r',z_\ell';0,1)),
\end{aligned} \tag{3.25}$$

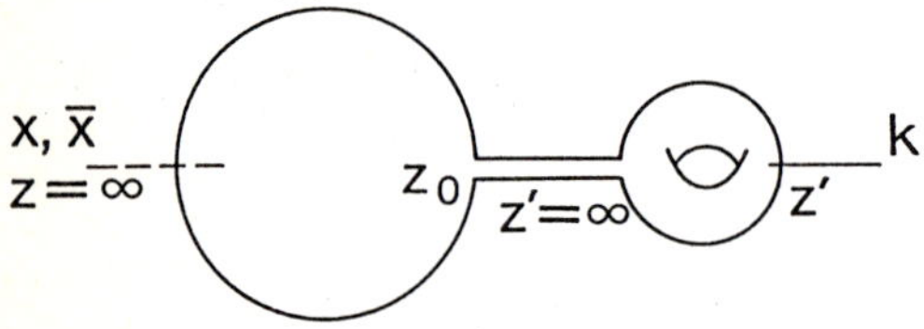

Fig. 3.3 The one-loop N=1 term gives a self-energy contribution. The conformal symmetry breaking occurs at z_0 due to the small handle with a single vertex operator insertion.

where we changed variables z_1' and z_2' to the rescaled variables z_1' and z_2' and used the SL(2, C) invariance of $\mathcal{G}$ in eq.(3.18). In performing the z_∞ integration, we observe that power divergences may occur depending on the value of k_1k_2. In order to exhibit essential features in a simpler context, let us consider an integral convergent for $\alpha>0$

$$I(\alpha) = \int_0^\infty dx\, x^{\alpha-1} f(x). \tag{3.26}$$

Since $I(\alpha)$ diverges for $-1<\alpha\leq 0$, we regularize the integral by a cut-off ε and separate the power divergence by a partial integration

$$\begin{aligned} \int_\varepsilon^\infty dx\, x^{\alpha-1} f(x) &= \frac{1}{\alpha}\varepsilon^\alpha f(0) + \tilde{I}(\alpha) + O(\varepsilon^{\alpha+1}), \\ \tilde{I}(\alpha) &= -\frac{1}{\alpha}\int_0^\infty dx\, x^\alpha f'(x), \end{aligned} \tag{3.27}$$

where $\tilde{I}(\alpha)$ in $-1<\alpha<0$ is identical to the analytic continuation of $I(\alpha)$ from $\alpha>0$ to $-1<\alpha<0$. One should note that the above formula (3.27) is valid for any α larger than -1. When $\alpha<-1$, we need to retain terms of order $\varepsilon^{\alpha+1}$.

Applying the above procedure for extracting power divergences when $k_1k_2>-2$, the integral (3.25) is separated into I_a and I_b corresponding to the first and the second terms in eq.(3.27)

$$\begin{aligned} I &= I_a + I_b \\ I_a &= \pi \ln\frac{1}{\varepsilon^2} \frac{-2\pi}{2k_1k_2+2}\varepsilon^{2k_1k_2+2}\int d^2z_1'\, e^{k^2\mathcal{G}(z_1')} \\ I_b &= \frac{-2\pi}{2k_1k_2+2}\varepsilon^{2k_1k_2+2}\int d^2z_1'\, d^2z_2'\, |z_1'-z_2'|^{2k_1k_2} e^{-\Sigma k_r k_\ell \mathcal{G}}, \end{aligned} \tag{3.28}$$

where the logarithmic divergence factor in I_a is due to the z_2' integral. If we assume $k_1k_2\leq -2$, a massless particle vertex operator generated by operator product expansion of tachyon vertex operators becomes a marginal or relevant operator ($(k_1+k_2)^2 \lesssim 0$ for $k_1^2 \simeq k_2^2 \simeq 2$). In that case, massless particle contributions should be taken into account and one finds coupled renormalization group equations. Since we are interested in the equation of motion for tachyon, we assume $k_1k_2>-2$ and obtain the second order terms in eq.(3.11) as a sum of "self-energy term" and "vertex term" corresponding to I_a and I_b respectively

$$\begin{aligned}\tau^{N=2}_{h=1}(x) &= \tau^{N=2}_{h=1,\mathrm{self}} + \tau^{N=2}_{h=1,\mathrm{vertex}}\\ \tau^{N=2}_{h=1,\mathrm{self}} &= \frac{1}{2}\int dk_1 dk_2 \phi(k_1)\phi(k_2)\int d^2z\,\varepsilon^{k^2-2}e^{ik\mathcal{X}(z)}\\ &\times \frac{-2\pi}{2k_1k_2+2}\cdot\frac{1}{2}\,\ln\frac{1}{\varepsilon^2}\,T_{h=1}(-k,k)\end{aligned} \tag{3.29}$$

$$\begin{aligned}\tau^{N=2}_{h=1,\mathrm{vertex}} &= \frac{1}{2}\int dk_1 dk_2 \phi(k_1)\phi(k_2)\int d^2z\,\varepsilon^{k^2-2}e^{ik\mathcal{X}(z)}\\ &\times \frac{-2\pi}{2k_1k_2+2}\,T_{h=1}(-k,k_1,k_2)\,,\end{aligned} \tag{3.30}$$

where $k=k_1+k_2$, the one-loop two-point function $T_{h=1}(-k,\ k)$ is given in eq.(3.24) and the one-loop N+1-point function $T_{h=1}(-k,\ k_1,\ \cdots,\ k_N)$ is defined as

$$\begin{aligned}T_{h=1}(-k,k_1,\cdots,k_N) &= (2\pi)^2\left(\frac{\kappa}{\pi}\right)^2\int_F \frac{d^2\tau}{(\mathrm{Im}\tau)^2}Z(\tau,\bar{\tau})\,\mathrm{Im}\tau\int\prod_{i=1} d^2z_i'\\ &\prod_{1\le i<j\le N}|z_i'-z_j'|^{2k_ik_j}\exp\Big(-\sum_{1\le r<\ell\le N+1}k_rk_\ell\,\mathcal{G}(z_r',z_\ell';0,1)\Big).\end{aligned} \tag{3.31}$$

The self-energy and vertex terms can be visualized as in Fig. 3.4 and 3.5 respectively.

Similarly we obtain the N-th order term in eq.(3.11) consisting of the self-energy and the vertex terms

$$\tau^{N}_{h=1}(x) = \tau^{N}_{h=1,\mathrm{self}} + \tau^{N}_{h=1,\mathrm{vertex}}\ .$$

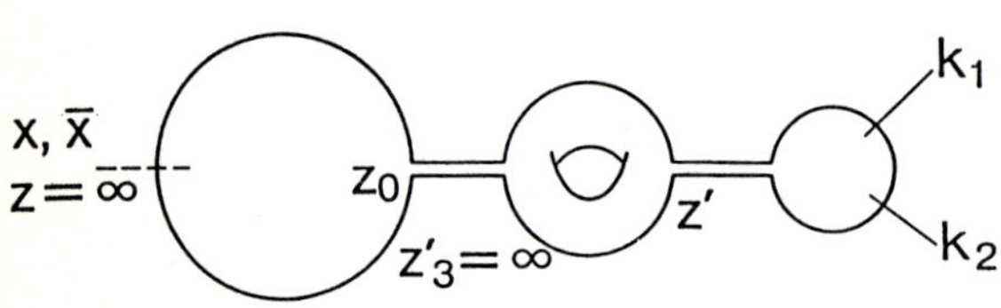

Fig. 3.4 The self-energy term (3.29) of the one-loop N=2 term. The logarithmic and power divergences occur due to the pinching of two homologically trivial cycles.

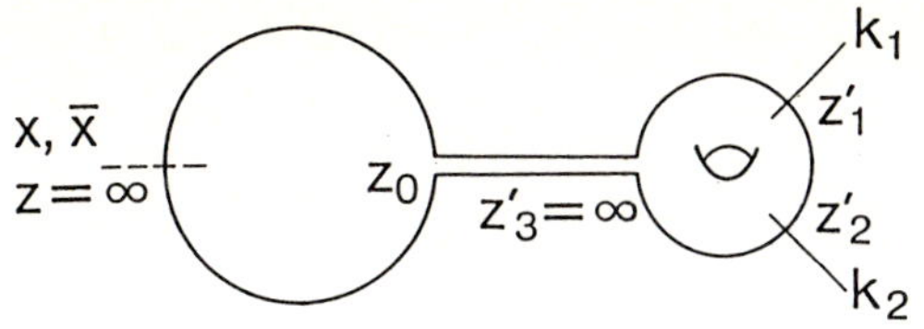

Fig. 3.5 The vertex term (3.30) of the one-loop N=2 term. The conformal symmetry breaking occurs at z_0 due to the small handle with two vertex operator insertions.

The self-energy term is given by

$$\tau^{N}_{h=1,\mathrm{self}} = \frac{1}{N!}\int dk \prod_{i=1}^{N} dk_i \phi(k_i)\delta(k-\Sigma k_i)\int d^2z\varepsilon^{k^2-2} e^{ik\mathfrak{X}(z)}$$
$$\times \frac{-2\pi}{2(\sum_{1\le i<j\le N} k_i k_j + N-1)} \frac{1}{2}\ln\frac{1}{\varepsilon^2} T_{h=1}(-k,k)\, t_0^{(N+1)}(-k,k_1,\cdots,k_N), \qquad (3.32)$$

where $t_0^{(N+1)}$ is a dual N+1 point function at the tree level defined in eq.(2.23). The vertex term is given in terms of the one-loop N+1 point function defined in eq.(3.31)

$$\tau^{N}_{h=1,\mathrm{vertex}} = \frac{1}{N!}\int dk \prod_{i=1}^{N} dk_i \phi(k_i)\delta(k-\Sigma k_i)\int d^2z\varepsilon^{k^2-2} e^{ik\mathfrak{X}(z)}$$
$$\frac{-2\pi}{2(\sum_{1\le i<j\le N} k_i k_j + N-1)} T_{h=1}(-k,k_1,\cdots,k_N)\ . \qquad (3.33)$$

Since the Schottky parametrization uses the two-sphere with holes to represent the torus, it is natural to cancel the short-distance divergences of $\tau_{h=0}$ and $\tau_{h=1}$ at the same coordinate point z. Physically this amounts to balance the conformal symmetry breakings due to the small handle at z against those due to the vertex operator insertions at the same z. By applying the renormalization group equation of the previous section to the τ-function $\tau(x)=\tau_{h=0}(x)+\tau_{h=1}(x)+\cdots$, we find the loop-corrected equations of motion for tachyons as the fixed point condition

$$0 = 2\beta(k) = (-k^2+2+T_{h=1}(-k,k))\phi(k)$$
$$+ \frac{1}{2!}\int dk_1 dk_2 \qquad \delta(k-k_1-k_2)\{2\pi + T_{h=1}(-k,k_1,k_2) - \frac{2\pi}{k^2-2}T_{h=1}(-k,k)$$

$$+\frac{2\pi}{2k_1k_2+2}(T_{h=1}(-k_1,k_1)+T_{h=1}(-k_2,k_2))\}\,\phi(k_1)\phi(k_2)+\cdots \tag{3.34}$$

where the last two terms become the self-energy subtraction terms in k_1 and k_2 channels. Similarly to the tree level β-function, we are thus able to obtain the loop corrected string equations of motion with one-particle irreducible interaction kernels at least up to the order exhibited above.

We observe that the logarithmic divergences associated with the self-energy terms in eqs.(3.23),(3.29) and (3.32) can be combined with the tree level (h=0) τ-function to give the first two terms of the expansion of the anomalous dimension exponent

$$\varepsilon^{k^2-2-T_{h=1}(-k,k)} = \varepsilon^{k^2-2}(1+\frac{1}{2}\,\ln\frac{1}{\varepsilon^2}\,T_{h=1}(-k,k)+\cdots).$$

Therefore we obtain to this order

$$\begin{aligned}\tau_\infty(x) &= \tau_{h=0}(x)+\tau_{h=1}(x)+\cdots\\ &= 1+\int dk \int d^2z\,\varepsilon^{k^2-2-T_{h=1}(-k,k)}\,e^{ik\mathcal{X}(z)}\\ &\qquad\times[W_{h=0}(\phi,k)+W_{h=1}(\phi,k)+\cdots]\end{aligned} \tag{3.35}$$

$$W_{h=1}(\phi,k)=\sum_{N=1}^{\infty}\frac{1}{N!}\int\prod_{i=1}^{N}dk_i\phi(k_i)\delta(k-\sum_{i=1}^{N}k_i)\cdot \cdot\frac{-1}{2(\sum_{1\le i<j\le N}k_ik_j+N-1)}T_{h=1}(-k,k_1,\cdots,k_N).$$

So far we have exhibited the divergence due to integration over handle positions z_0, z_∞ and vertex positions z_1, $\cdots$, z_N. Bosonic string contains an extra divergence from the integration over τ due to the presence of tachyon. We shall explain in ref.[47] the reason why the divergence due to τ integration should not be counted as a source of the conformal symmetry breaking in obtaining the equations of motion of string.

We notice that the one-loop amplitude $T_{h=1}(-k, k_1, \cdots, k_N)$ for N>1 still contains the divergence due to the tadpole. Since the tadpole should be absent on a stable vacuum as in the case of superstring, we have not made efforts to disentangle the tadpole divergence completely.

3.2 Nonlinear sigma model on the genus h surfaces

The n-point h-loop amplitudes are given by a path integral on a Riemann surface with h-handles and n punctures. The number of moduli parameters of the surface with h-handles is given by 3h-3 for $h \geq 2$, one for h=1, and zero for h=0. There are N independent vertex positions for $h \geq 2$, but there are only N-1 for h=1 and N-3 for h=0 because of the translation and SL(2, C) invariances on h=1 and h=0 respectively. Therefore the total number of independent integration variables is N-3+3h. The genus h Riemann surface can be represented as a two-sphere with h pairs of holes cut out and identified pairwise as shown in Fig. 3.6. This parametrization is called the Schottky parametrization. The identification of the s-th pair of holes can be described by an SL(2, C) transformation T_s with three complex parameters z_{0s}, $z_{\infty s}$, q_s

$$\frac{z'-z_{0s}}{z'-z_{\infty s}} = q_s\frac{z-z_{0s}}{z-z_{\infty s}}, \qquad q_s = e^{2\pi i\tau_s}, \qquad |q_s|<1 . \tag{3.36}$$

This transformation maps the entire region outside of the two holes into a region inside the circle enclosing the point z_{0s}. The two limiting points z_{0s} and $z_{\infty s}$ are fixed by T_s. Intuitively z_{0s} and $z_{\infty s}$

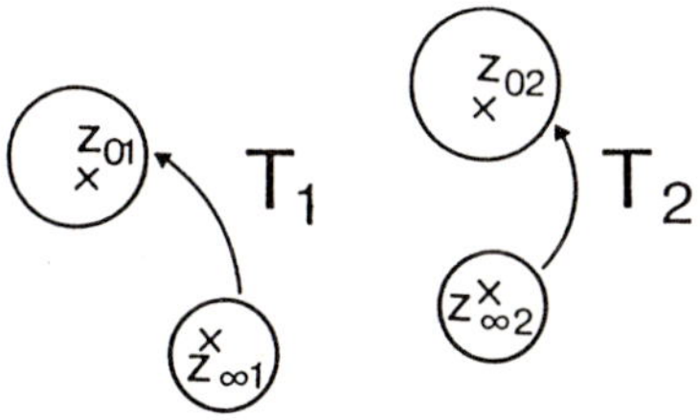

Fig. 3.6 The Schottky parametrization for a Riemann surface. Holes around the limiting points z_{0s} and $z_{\infty s}$ are cut out and identified pairwise by the SL(2, C) transformation T_s.

correspond to two positions at which the s-th handle is attached to the sphere and q_s represents the length and twist of the handle as illustrated in Fig. 3.1.

Let us summarize the Mandelstam result [27, 28] in our notation. The Abelian differential $\omega_s = dv_s$ $(s=1, \cdots, h)$ is given as an infinite sum over the group of SL(2, C) transformations V_i (the Schottky group G) generated by T_s in eq.(3.36)

$$v_s(z) = \frac{1}{2\pi i} \sum_i^{(s)} \ln\frac{z-V_i z_{0s}}{z-V_i z_{\infty s}} , \tag{3.37}$$

where $\Sigma^{(s)}$ denotes a summation over all values of i except those values for which V_i has a factor T_s or T_s^{-1} at its right-hand end. The rs element of the period matrix Ω is given by

$$\begin{aligned} \Omega_{rs} &= \frac{1}{2\pi i} \sum_i^{(r,s)} \ln\frac{(z_{0s}-V_i z_{0r})(z_{\infty s}-V_i z_{\infty r})}{(z_{0s}-V_i z_{\infty r})(z_{\infty s}-V_i z_{0r})} , \qquad r\neq s \\ \Omega_{rr} &= \frac{1}{2\pi i} \{\ln q_r + \sum_{i\neq I}^{(r,r)} \ln\frac{(z_{0r}-V_i z_{0r})(z_{\infty r}-V_i z_{\infty r})}{(z_{0r}-V_i z_{\infty r})(z_{\infty r}-V_i z_{0r})}\} , \end{aligned} \tag{3.38}$$

where the summation $\Sigma^{(r,s)}$ excludes values of r and s such that V_i has T_r^{+1} as its rightmost member or T_s^{+1} as its leftmost member. The scalar field Green's function is given for the genus h surface in the Schottky parametrization

$$\langle X(z)X(w)\rangle = -\ln|\phi(z,w)|^2 = -\ln|z-w|^2 + \mathcal{G}(z,w;z_{01},z_{\infty 1};\cdots;z_{0h},z_{\infty h}) . \tag{3.39}$$

The function ϕ may be called a modified prime form and is given by

$$\begin{aligned} \ln\phi(z,w) &= \ln(z-w) + \sum_{i=I}' \ln\frac{(z-V_i w)(w-V_i z)}{(z-V_i z)(w-V_i w)} \\ &- \pi \sum_{r,s} \mathrm{Im}(v_r(z)-v_r(w))\left(\frac{1}{\mathrm{Im}\Omega}\right)_{r,s} \mathrm{Im}(v_s(z)-v_s(w)) , \end{aligned} \tag{3.40}$$

where the summation Σ' counts V_i and V_i^{-1} together only once. We note that the function $\mathcal{G}$ is SL(2, C) invariant. Mandelstam has given a formula for the h-loop n-tachyon amplitudes in the Schottky parametrization which reads in our convention [27, 28]

$$A_n(1,\cdots,n) = \frac{2\pi}{(4\pi(8\pi^2)^{13})^h}\left(\frac{\kappa}{\pi}\right)^{n+2h-2}\int\frac{\prod_{i=1} d^2z_i}{V(SL(2,C))}$$

$$\cdot\int\prod_{s=1}^{h}\frac{d^2z_{0s}d^2z_{\infty s}d^2q_s}{|q_s(z_{0s}-z_{\infty s})|^4}\frac{\prod'_i\prod_{n\geq 1}|1-w_i{}^n|^{-48}}{(\det \mathrm{Im}\Omega)^{13}}\prod_{1<i<j<n}|\phi(z_i-z_j)|^{2k_ik_j}\cdot\Gamma,$$

$$\Gamma = \prod_{s=1}^{h}|1-q_s|^4 \;/\; \prod'_i|1-w_i|^4 , \tag{3.41}$$

where the infinite product $\prod'_i \prod_{n=1}^{\infty} |1-w_i{}^n|^{-48}$ corresponds to $\prod_{n\geq 1}(1-e^{2\pi in\tau})$ in the one-loop case and w_i is the multiplier of SL(2, C) transformations in the Schottky group (corresponding to q_s in eq.(3.36)). The product $\prod'_i$ is the product over all primitive elements of the Schottky group, i.e. those that cannot be written as powers of other elements, with each conjugacy class counted only once. The factor Γ is the correction to the Mandelstam's formula pointed out by Di Vecchia et al. [28].

As described in Appendix B, we can construct a h-handle operator e^G which enables us to express correlation functions on the genus h surface as an expectation value of corresponding operators defined on two-sphere

$$<\prod_{j=1}^{n} V_j(z_j)> = <0|\prod_{j=1}^{n}\hat{V}_j(z_j)\; e^G|0> . \tag{3.42}$$

For tachyon vertex operators the correlation function is given by the modified prime form

$$<0|\prod_{j=1}^{n} :e^{ik\hat{X}(z_j)}:e^G|0> = \prod_{1\leq i<j\leq n}|\phi(z_i,z_j)|^{2k_ik_j}. \tag{3.43}$$

$$= \prod_{1\leq i<j\leq n}|\, z_i-z_j\, |^{2k_ik_j}\exp(-\sum_{1\leq i<j\leq n} k_ik_j\mathcal{G}(z_i,z_j;z_{01},z_{\infty 1};\cdots;z_{0h},z_{\infty h})) .$$

Similarly to the h=1 case in sect. 2, the nonlinear sigma model can be defined on genus h surfaces. We can probe the nonlinear sigma model by inserting a finite number of vertex operators near a point on the surface whose coordinates in the Schottky parameterization are near infinity. Then we can factorize the genus h vacuum state $|\phi>_{(h)}$ with the background

$$|\phi>_{(h)} = \frac{1}{(4\pi(8\pi^2)^{13})^h}\left(\frac{\kappa}{\pi}\right)^{2h}\int\prod_{s=1}^{h}\frac{d^2z_{0s}d^2z_{\infty s}d^2q_s}{|q_s(z_{0s}-z_{\infty s})|^4}$$

$$(\det \mathrm{Im}\Omega)^{-13}\prod_i{}' \prod_{n\geq 1}|1-w_i{}^n|^{-48}\Gamma\, e^{\hat{S}_{int}} e^{G}|0>. \tag{3.44}$$

It is natural to define the vacuum state vector $|\phi>$ of the nonlinear sigma model on the sum over surfaces with various genera and the corresponding τ-function

$$|\phi>_\infty = \sum_{h=0}^{\infty}|\phi>_{(h)},$$

$$\tau_\infty(x) = \sum_{h=0}^{\infty}\tau_h(x) = \sum_{h=0}^{\infty}<x,\bar{x}|\phi>_{(h)}. \tag{3.45}$$

The factorization property of string amplitudes can be exhibited from degenerate Riemann surfaces obtained by pinching various cycles [45]. When we pinch the s-th trivial cycle of the genus h surface, we obtain genus s-1 and h-s+1 surfaces connected by a node as illustrated in Fig. 3.7. This pinching is realized in the Schottky parametrization as shrinking the genus h-s+1 surface: $z_{\infty s}\to z_{0s}$, fixing the rescaled variables $z'_{0r}=(z_{0r}-z_{0s})(z_{\infty s}-z_{0s})^{-1}$, $z'_{\infty r}=(z_{\infty r}-z_{0s})(z_{\infty s}-z_{0s})^{-1}$ for r=s+1, ..., h together with z_{0r}, $z_{\infty r}$, r=1, ..., s-1. The Abelian integral for the r-th cycle of the genus h

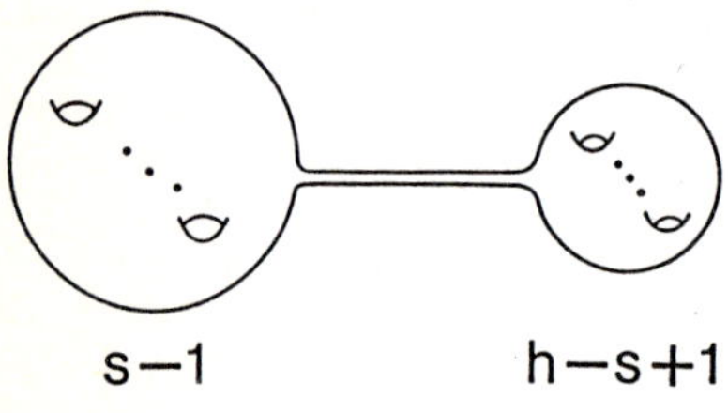

Fig. 3.7 Pinching of a homologically trivial cycle of a genus h surface. Surfaces with the genus s-1 and h-s+1 are connected by a node.

surface reduces to that on the genus s-1 or h-s+1 surface in this limit:

for fixed z

$$v_r^{(h)}(z) \longrightarrow \begin{cases} v_r^{(s-1)}(z) & r=1,\cdots,s-1 \\ 0 & r=s,\cdots,h \end{cases} \tag{3.46}$$

for $z \to z_{0s}$ with $z'=(z-z_{0s})/(z_{\infty s}-z_{0s})$ fixed

$$v_r^{(h)}(z) \longrightarrow \begin{cases} 0 & r=1,\cdots,s-1 \\ v_r^{(h-s+1)}(z) & r=s,\cdots,h \end{cases} .$$

The period matrix also becomes diagonal

$$\Omega_{r\ell}^{(h)} \longrightarrow \begin{cases} \Omega_{r\ell}^{(s-1)} & r,\ell = 1, \cdots, s-1 \\ \Omega_{r\ell}^{(h-s+1)} & r,\ell = s, \cdots, h \\ 0 & \text{otherwise}. \end{cases} \tag{3.47}$$

The $\mathcal{G}$-function appeared in the scalar Green's function (3.39) which also shows a factorization property [46]: Depending on three cases as illustrated in Fig. 3.8 a), b) and c)

a) $z-z_{0s}$, $w-z_{0s}$ fixed $\quad (|z-z_{0s}|, |w-z_{0s}| >> |z_{\infty s}-z_{0s}|)$

$$\mathcal{G}(z,w;z_{01},z_{\infty 1};\cdots;z_{0h},z_{\infty h}) \longrightarrow \mathcal{G}(z,w;z_{01},z_{\infty 1};\cdots;z_{0s-1},z_{\infty s-1})$$

b) $z' = (z-z_{0s})/(z_{\infty s}-z_{0s})$, $\quad w' = (w-z_{0s})/(z_{\infty s}-z_{0s})$ fixed $\quad$ (3.48)

$$\mathcal{G}(z,w;z_{01},z_{\infty 1};\cdots;z_{0h},z_{\infty h}) \longrightarrow \mathcal{G}(z,w;z_{0s},z_{\infty s};\cdots;z_{0h},z_{\infty h})$$

c) $z-z_{0s}$, w' fixed

$$\mathcal{G}(z,w;z_{01},z_{\infty 1};\cdots;z_{0h},z_{\infty h}) \longrightarrow \mathcal{G}(z,z_{0s};z_{01},z_{\infty 1};\cdots;z_{0s-1},z_{\infty s-1}) + \mathcal{G}(\infty,w';z'_{0s},z'_{\infty s};\cdots;z'_{0h},z'_{\infty h}) \tag{3.49}$$

The infinite product of conjugacy classes factorizes*)

$$[\prod_i{}' \prod_{n\geq 1} (1-w_i^{\,n})]^{(h)} \longrightarrow [\prod_i{}' \prod_{n\geq 1} (1-w_i^{\,n})]^{(s-1)} \cdot [\prod_i{}' \prod_{n\geq 1} (1-w_i^{\,n})]^{(h-s+1)} ,$$

---

*) This property is necessary to verify the factorization and hence unitarity of string amplitude in the Schottky parametrization, although we have not yet explicitly demonstrated it.

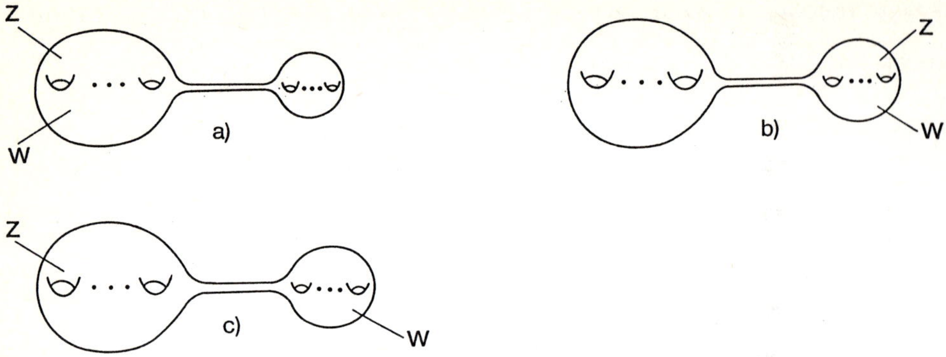

Fig. 3.8 The factorization of the scalar Green's function (prime form) in the Schottky parametrization.

a) When two scalar fields are on the genus s-1 surface.

b) When two scalar fields are on the genus h-s+1 surface.

c) When one scalar field at z is on the genus h-1 surface and the other at w is on the h-s+1 surface.

$$\Gamma^{(h)} \longrightarrow \Gamma^{(s-1)} \cdot \Gamma^{(h-s+1)}. \tag{3.50}$$

On the other hand we find in the coincidence limit

$$\mathcal{G}(z,w;z_{01},z_{\infty 1};\cdots;z_{0h},z_{\infty h}) \longrightarrow 0 \text{ as } z \to w. \tag{3.51}$$

3.3 Two-loop case (genus two)

Let us examine the short-distance divergence of the nonlinear sigma model on the genus two (h=2) surface in the weak field expansion

$$\tau_{h=2}(x) = \sum_{N=0}^{\infty} \tau^{N}_{h=2}(x). \tag{3.52}$$

Using the general formula (3.44) and a procedure similar to sect. 3.1, we find the short-distance divergence of the form $e^{ik\mathfrak{X}(z_{01})}$ as

$$\tau^{N}_{h=2}(x) = \frac{1}{(4\pi(8\pi^2)^{13})^2}\left(\frac{\kappa}{\pi}\right)^4 \int \frac{d^2q_1 d^2q_2}{|q_1q_2|^4}\, \frac{\prod''_{i=I}|1-w_i|^{-48}}{(\det \mathrm{Im}\Omega)^{13}} \cdot \Gamma$$

$$\times \int \prod_{i=1}^{N} dk_i \phi(k_i)\ \bar{\tau}^{N}_{h=2}$$

$$\bar{\tau}^{N}_{h=2} = \frac{1}{N!}\int \prod_{i=1}^{N} d^2 z_i \varepsilon^{k_i^2-2} \prod_{s=1}^{2} \frac{d^2 z_{0s} d^2 z_{\infty s}}{(|z_{0s}-z_{\infty s}|^2+\varepsilon^2)^2}\ e^{i\sum_{i=1}^{N} k_i \mathcal{X}(z_{01})} \tag{3.53}$$

$$\prod_{1\le i<j\le N} (|z_i-z_j|^2+\varepsilon^2)^{k_i k_j} \exp(-\sum_{1\le r<\ell\le N+1} k_r k_\ell \mathcal{G}(z_r, z_\ell; z_{01}, z_{\infty 1}; z_{02} z_{\infty 2})),$$

where we introduced the cut-off ε^2 as in eq.(2.8), and $z_{N+1}=\infty$ and $k_{N+1}=-(k_1+\cdots k_N)\equiv -k$ as in sect. 3.1. It is convenient to rescale variables

$$z' = (z - z_{01})(z_{\infty 1} - z_{01})^{-1} \tag{3.54}$$

for $z=z_1, \cdots, z_N, z_{02}$ and $z_{\infty 2}$. Since the function $\mathcal{G}$ is SL(2, C) invariant, we find

$$\mathcal{G}(z_r, z_\ell; z_{01}, z_{\infty 1}; z_{02}, z_{\infty 2}) = \mathcal{G}(z_r', z_\ell'; 0, 1; z_{02}, z_{\infty 2}). \tag{3.55}$$

We also use further rescaled variables

$$\begin{aligned} z_i'' &= (z_i - z_{02})(z_{\infty 2} - z_{02})^{-1}, \\ z_i''' &= (z_i - z_1)(z_2 - z_1)^{-1}, \qquad i=1,\cdots,N \end{aligned} \tag{3.56}$$

to find the short-distance divergence for $N\ge 2$

$$\bar{\tau}^{N}_{h=2} = \frac{1}{N!}\int d^2 z_{01} \varepsilon^{\sum_i (k_i^2 -2)} e^{ik\mathcal{X}(z_{01})} \times I \tag{3.57}$$

$$I = \int d^2 z_{\infty 1} (|z_{\infty 1}-z_{01}|^2+\varepsilon^2)^{\sum_{1\le i<j\le N} k_i k_j + N-2}$$
$$\times \int d^2 z'_{02} d^2 z'_{\infty 2} (|z'_{\infty 2}-z'_{02}|^2 + \frac{\varepsilon^2}{|z_{\infty 1}-z_{01}|^2})^{\Sigma k_i k_j + N-2} \times J$$

$$J = \int d^2 z_1'' d^2 z_2'' (|z_2''-z_1''|^2 + \frac{\varepsilon^2}{|z'_{\infty 2}-z'_{02}|^2\ |z_{\infty 1}-z_{01}|^2})^{\Sigma k_i k_j + N-2} \tag{3.58}$$
$$\times \int \prod_{i=3}^{N} d^2 z_i'' \prod_{1\le i<j\le N} |z_i''-z_j''|^{2k_i k_j} \exp(-\sum_{1\le r<\ell\le N+1} \mathcal{G}\, k_r k_\ell (z'_r, z'_\ell, 0, 1; z'_{02} z'_{\infty 2})).$$

Since we are interested in the short-distance divergence associated to the tachyon in the $k=k_1+\cdots+k_N$ channel, we assume $\sum_{1\le i<j\le N} k_ik_j+N-2\gtrsim-2$, but $k_ik_j>-1$. Hence we do not need a cut-off at singularities $z_i'''\sim z_j'''$. After a partial integration in z_2'' and using eq.(3.50) we obtain

$$J \simeq J_a + J_b$$

$$J_a = \frac{-2\pi}{2(\Sigma k_ik_j+N-1)}\left(\frac{\varepsilon^2}{|z'_{\infty2}-z'_{02}|^2\ |z_{\infty1}-z_{01}|^2}\right)^{\Sigma k_ik_j+N-1}\int d^2z_1''$$
$$\times\ e^{k^2\mathcal{G}(z_1',\infty;0,1;z'_{02}z'_{\infty2})}\int\prod_{i=3}^{N}d^2z_i''\prod_{1\le i<j\le N}|z_i''-z_j''|^{2k_ik_j} \tag{3.59}$$

$$J_b = \int d^2z_1'''d^2z_2'''|z_2'''-z_1'''|^{2(\Sigma k_ik_j+N-2)}\int\prod_{i=3}^{N}d^2z_i''\prod_{1\le i<j\le N}|z_i''-z_j''|^{2k_ik_j}$$
$$\times\ \exp\Bigl(-\sum_{1\le r<\ell\le N+1}k_rk_\ell\,\mathcal{G}(z_r',z_\ell',0,1;z'_{02}z'_{\infty2})\Bigr), \tag{3.60}$$

where J_b is obtained as an analytic continuation of the z_2'' integral defined in $\Sigma k_ik_j+N-2>-1$.

Let us first examine the contribution of J_b to the integral I in eq.(3.57), which we denote as I_b. A partial integration in $z'_{\infty2}$ gives two terms for I_b

$$I_b = I_{ba} + I_{bb}$$

$$I_{ba} = \int\frac{d^2z_{\infty1}}{|z_{\infty1}-z_{01}|^2+\varepsilon^2}\,\frac{-2\pi}{2(\Sigma k_ik_j+N-1)}\,\varepsilon^{2(\Sigma k_ik_j+N-1)}$$
$$\times\int d^2z'_{02}\prod_{i=1}^{N}d^2z_i''\prod_{1\le i<j\le N}|z_i''-z_j''|^{2k_ik_j}\,e^{-\Sigma k_rk_\ell\mathcal{G}}, \tag{3.61}$$

$$I_{bb} = \int d^2z_{\infty1}(|z_{\infty1}-z_{01}|^2+\varepsilon^2)^{\Sigma k_ik_j+N-2}$$
$$\times\int d^2z'_{02}d^2z'_{\infty2}(|z'_{\infty2}-z'_{\infty1}|^{2(\Sigma k_ik_j+N-2)}$$
$$\times\int\prod_{i=1}^{N}d^2z_i''\prod_{1\le i<j\le N}|z_i''-z_j''|^{2k_ik_j}\,e^{-\Sigma k_rk_\ell\mathcal{G}}, \tag{3.62}$$

where I_{bb} is obtained as an analytic continuation of the $z'_{\infty2}$ integral defined in $\Sigma k_ik_j+N-2>-1$.

The integral I_{ba} has a short-distance divergence as $z_{\infty1}\to z_{01}$ with z'_{02}, z_1'', $\cdots$, z_N'' fixed. Because of the factorization property in eqs.(3.48) and (3.49), we obtain the exponent with $\mathcal{G}$-function in this limit as a sum of the one-loop $\mathcal{G}$-functions

$$
\begin{aligned}
& - \sum_{1\le r<\ell\le N+1} k_r k_\ell \mathcal{G}(z'_r, z'_\ell; z_{01}, z_{\infty 1}; z'_{02} z'_{\infty 2}) \\
& \longrightarrow k^2 \mathcal{G}^{\tau_1}(z'_{02}, \infty; 0,1) - \sum_{1\le r<\ell\le N+1} k_r k_\ell \mathcal{G}^{\tau_2}(z''_r, z''_\ell; 0,1)
\end{aligned}
\tag{3.63}
$$

with $k_{N+1}=-k=-\sum_{i=1}^{N} k_i$ and $z''_{N+1}=\infty$. Moreover, the factorization property (3.47) and (3.50) implies in the limit

$$
\begin{aligned}
& \det(\mathrm{Im}\Omega) \longrightarrow \mathrm{Im}\tau_1 \cdot \mathrm{Im}\tau_2, \\
& \prod_i{}' \prod_{n\ge1} (1-w_i{}^n) \cdot \Gamma \longrightarrow \prod_{n\ge1} |1-e^{2\pi i n\tau_1}| \cdot |1-e^{2\pi i n\tau_2}|.
\end{aligned}
\tag{3.64}
$$

Hence the contribution to $\tau^N_{h=1}$ from I_{ba} is given as a product of the one-loop self-energy and the one-loop vertex amplitudes

$$
\begin{aligned}
& \tau^N_{h=2}\Big|_{ba} = \frac{1}{N!}\int dk \prod_{i=1}^{N} dk_i \phi(k_i)\delta(k-\Sigma k_i)\int d^2 z \varepsilon^{k^2-2} e^{ik\mathcal{X}(z)} \\
& \frac{-1}{2(\Sigma k_i k_j + N-1)} \frac{1}{2} \ln\frac{1}{\varepsilon^2} T_{h=1}(-k,k) T_{h=1}(-k,k_1,\cdots,k_N),
\end{aligned}
\tag{3.65}
$$

where the one-loop amplitudes are given in eqs.(3.24) and (3.31). This contribution from I_{ba} can be visualized as in Fig. 3.9. We shall find that this contribution enables us to exponentiate the self-energy as a shift of the anomalous dimension.

Another integral I_{bb} in eq.(3.62) has a short-distance divergence due to $z_{\infty 1} \to z_{01}$ with z'_{02}, $z'_{\infty 2}$, z'_1, $\cdots$, z'_N fixed

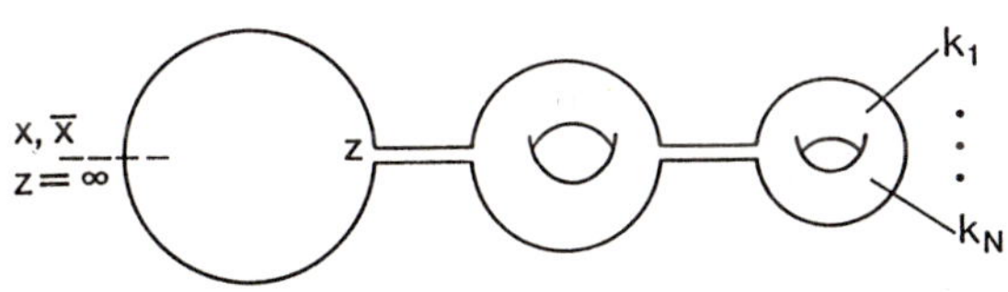

Fig. 3.9 The product of the one-loop self-energy and the one-loop vertex amplitudes as a short-distance divergence of the two-loop N-th order term.

$$I_{bb} \sim \frac{-2\pi}{2(\Sigma k_i k_j+N-1)} \varepsilon^{2(\Sigma k_i k_j+N-1)} \int \frac{d^2 z'_{02} d^2 z'_{\infty 2}}{|z'_{\infty 2}-z'_{02}|^4} \prod_{i=1}^{N} d^2 z'_i$$

$$\prod_{1\le i<j\le N} |z''_i - z''_j|^{2k_i k_j} \exp(-\sum_{1\le r<\ell\le N+1} k_r k_\ell \mathcal{G}(z'_r, z'_\ell; 0,1; z'_{02} z'_{\infty 2})). \quad (3.66)$$

Therefore we obtain the part of $\tau^N_{h=2}$ coming from I_{bb} as

$$\tau^N_{h=2}\Big|_{bb} = \frac{1}{N!}\int dk \prod_{i=1}^{N} dk_i \phi(k_i) \delta(k-\Sigma k_i) \int d^2 z \varepsilon^{k^2-2} e^{ik\mathcal{X}(z)}$$

$$\cdot \frac{-1}{2(\Sigma k_i k_j+N-1)} T_{h=2}(-k, k_1, \cdots, k_N), \quad (3.67)$$

where the two-loop amplitude is defined as

$$T_{h=2}(-k,k_1,\cdots,k_N) = \frac{2\pi}{(4\pi(8\pi^2)^{13})^2} \left(\frac{\kappa}{\pi}\right)^4 \int \prod_{s=1}^{2} \frac{d^2 q_s}{|q_s|^4} \frac{1}{(\det \mathrm{Im}\Omega)^{13}}$$

$$\prod_i{}' \prod_{n\ge 1} |1-w_i{}^n|^{-48} \Gamma \int \frac{d^2 z_{\infty 2} d^2 z'_{02}}{|z'_{\infty 2}-z'_{02}|^4} \prod_{i=1}^{N} d^2 z'_i \prod_{1\le i<j\le N} |z'_i - z'_j|^{2k_i k_j}$$

$$\exp(-\sum_{1\le r<\ell\le N+1} k_r k_\ell \mathcal{G}(z'_r, z'_\ell; 0,1; z'_{02} z'_{\infty 2})). \quad (3.68)$$

One can recognize this two-loop amplitude as a gauge-fixed version ($z'_{01}=0$, $z'_{\infty 1}=1$, $z_{N+1}=\infty$) of the h=2, n=N+1 case of eq.(3.41). The contribution in eq.(3.67) can be visualized as a two-loop corrections to the N+1-point vertex in Fig. 3.10.

Next we evaluate the contribution from J_a in eq.(3.59) to the integral I in eq.(3.57), which we denote as I_a. We see immediately that the tree N+1-point function in eq.(2.23) can be factored out

$$\tau^N_{h=2}\Big|_a = \frac{1}{N!} \frac{-2\pi}{2(\Sigma k_i k_j+N-1)} t_0^{(N+1)}(-k, k_1, \cdots, k_N) \tau^{N=1}_{h=2} . \quad (3.69)$$

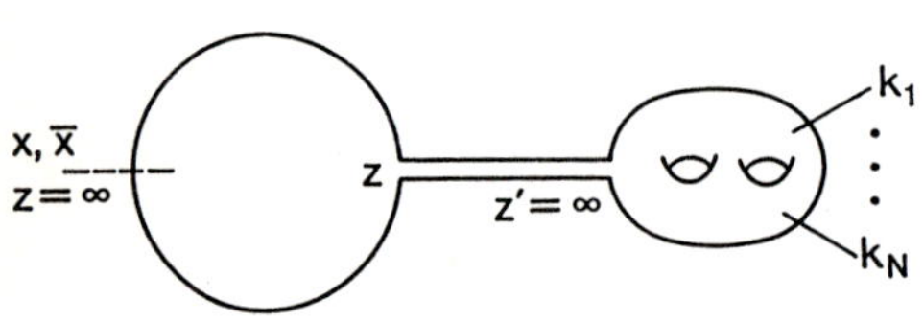

Fig. 3.10 The two-loop correction to the N+1 point vertex as a conformal symmetry breaking at z.

Hence we evaluate the N=1 case of eq.(3.53)

$$\tau^{N=1}_{h=2} = \int d^2z_{01}\varepsilon^{k^2-2}e^{ik\mathcal{X}(z_{01})}\int\frac{d^2z_{\infty1}}{|z_{\infty1}-z_{01}|^2+\varepsilon^2}$$

$$\int\frac{d^2z'_{02}d^2z'_{\infty2}d^2z'_1}{|z'_{02}-z'_{\infty2}|^2+\frac{\varepsilon^2}{|z_{01}-z_{\infty1}|^2}}\exp(k^2\mathcal{G}(z',\infty';0,1;z'_{02}z'_{\infty2})), \tag{3.70}$$

where we used the SL(2, C) invariance of the $\mathcal{G}$ -function. In performing the $z_{\infty1}$ integration, we find a logarithmic divergence. In order to illustrate essential features in extracting the logarithmic devergence, let us consider an integral regularized by a cut-off ε

$$\int_\varepsilon^\infty dx\,\frac{1}{x}\,f(x) = f(0)\,\ln\frac{1}{\varepsilon} - \int_0^\infty dx\,f(x)\,\ln x + O(\varepsilon). \tag{3.71}$$

The finite term can be expressed as a limit (an analytic regularization)

$$-\int_0^\infty dx\,f(x)\,\ln x = \lim_{\alpha\to0}\left(-\int_0^\infty dx\,x^{\alpha-1}f(x) - \frac{1}{\alpha}\,f(0)\right). \tag{3.72}$$

In order to interpret the pole $1/\alpha$ in the regularization parameter, we notice that the integration variable $z'_{\infty2}-z'_{02}$ represents the off-diagonal element of the period matrix Ω_{12}

$$\Omega_{12} \simeq \frac{1}{2\pi i}(z'_{\infty2}-z'_{02})\,z'^{\,-1}_{02}\,(z'_{02}-1)^{-1} \tag{3.73}$$

in the limit $z'_{\infty2} \to z'_{02}$ corresponding to the pinching of a trivial cycle as shown in Fig. 3.11. In the usual analysis of this pinching limit [45], n-point two-loop amplitudes possess a factor

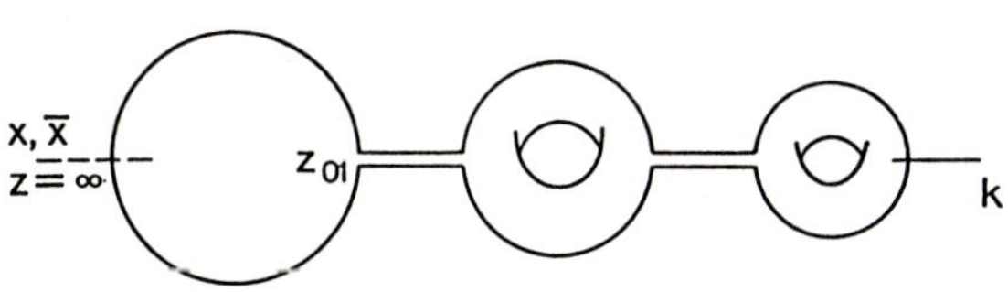

Fig. 3.11 The repetition of one-loop self-energy as the leading divergence of the two-loop N=1 term.

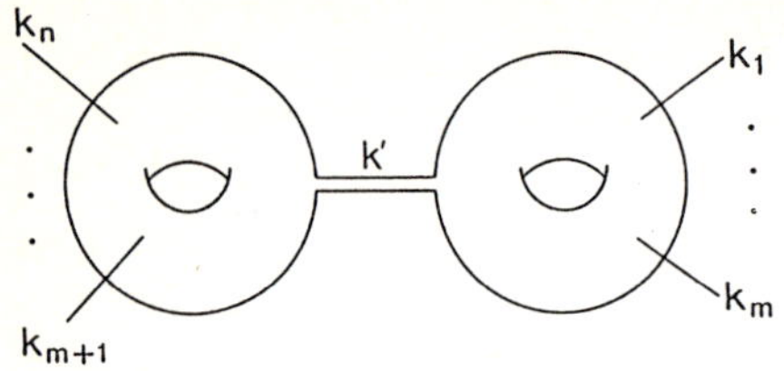

Fig. 3.12 The factorization of two-loop n-point amplitudes by pinching a homologically trivial cycle which divides n particles into m and n-m.

$$\frac{1}{2\pi}\int d^2\Omega_{12}\ |\Omega_{12}|^{k'^2-4} = (k'^2-2)^{-1} + \cdots$$

producing pole terms in the pinched channel (with momentum k') if two or more particles are factorized for each one-loop amplitudes: $k' = \sum_{i=1}^{m} k_i = -\sum_{i=m+1}^{n} k_i$, $m\geq 2$, $n-m\geq 2$ as shown in Fig. 3.12. In view of this fact, we can understand that the finite part of our τ-function is the on-shell limit of the one-tachyon-pole subtracted amplitude by identifying the regularization parameter α with k'^2-2. In the limit $z'_{\infty 2} \to z'_{02}$ dwith z''_1 fixed,   becomes a sum of one-loop $\mathcal{J}$ -functions

$$\mathcal{J}(z',\infty;0,1;z'_{02}z'_{\infty 2}) \longrightarrow \mathcal{J}(z'_{02}) + \mathcal{J}(z''_1). \tag{3.74}$$

Corresponding to two terms in eq.(3.71), we thus obtain two types of contributions

$$\begin{aligned}
\bar{\tau}^{N=1}_{h=2} &= \bar{\tau}|_{aa} + \bar{\tau}|_{ab},\\
\bar{\tau}|_{aa} &= \int d^2z_{01}\varepsilon^{k^2-2}e^{ik\mathcal{X}(z_{01})}\int d^2z'_{02}e^{k^2\mathcal{J}(z'_{02})}\int d^2z''_1 e^{k^2\mathcal{J}(z''_1)}\\
&\quad\cdot\int\frac{d^2z_{\infty 1}}{|z_{01}-z_{\infty 1}|^2+\varepsilon^2}\ \pi\ln\left(\frac{|z_{01}-z_{\infty 1}|^2}{\varepsilon^2}\right),\\
\bar{\tau}|_{ab} &= \int d^2z_{01}\varepsilon^{k^2-2}e^{ik\mathcal{X}(z_{01})}\int d^2z'_{02}d^2z''_1\int\frac{d^2z_{\infty 1}}{|z_{01}-z_{\infty 1}|^2+\varepsilon^2}\\
&\quad\cdot\lim_{k'^2\to 2}\left(\int d^2z'_{\infty 2}|z'_{\infty 2}-z'_{02}|^{k'^2-4}e^{k^2\mathcal{J}} - \frac{2\pi}{k'^2-2}e^{k^2(\mathcal{J}(z'_{01})+\mathcal{J}(z''_1))}\right),
\end{aligned}\tag{3.75}$$

where we have used the factorization property (3.74) in the pinching limit.

Thus we find the contribution to the τ-function from $\bar{\tau}|_{aa}$ as

$$\bar{\tau}^{N=1}_{h=2}|_{aa} = \int dk\phi(k)\int d^2z\varepsilon^{k^2-2}e^{ik\mathfrak{X}(z)}(T_{h=1}(-k,k))^2\ \frac{1}{8}(\log\frac{1}{\varepsilon^2})^2, \tag{3.76}$$

which can be visualized to be a repetition of one-loop self-energy corrections as shown in Fig. 3.11. This contribution is precisely the second order term needed to exponentiate the logarithmic term of the one-loop self-energy eq.(3.23). Hence we are led to interpret the self-energy diagram as a correction to the anomalous dimension.

Another term $\bar{\tau}|_{ab}$ contributes to the τ-function as

$$\begin{aligned}\tau^{N=1}_{h=2}|_{ab} = &\int dk\phi(k)\int d^2z\varepsilon^{k^2-2}e^{ik\mathfrak{X}(z)}\ \frac{1}{2}\log\frac{1}{\varepsilon^2} \\ &\times \lim_{k'^2\to 2}\{T_{h=2}(-k,k) - T_{h=1}(-k,k)\frac{1}{k'^2-2}\ T_{h=1}(-k,k)\}.\end{aligned} \tag{3.77}$$

One should note that the first term in the curly bracket depends on k'^2 and has a pole at $k'^2=2$ with the residue precisely equal to the second term. We interpret the above contribution as the self-energy correction to two-loop order with the one-particle reducible part subtracted, as illustrated in Fig. 3.13. Hence we can put all the self-energy corrections as a shift in the anomalous dimension in eq.(2.19)

$$\begin{aligned}&k^2-2 \longrightarrow k^2-2 - [T(-k,k)]_{1PI} \\ &[T(-k,k)]_{1PI} = T_{h=1}(-k,k) \\ &\qquad + [T_{h=2}(-k,k) - T_{h=1}(-k,k)\frac{1}{k'^2-2}\ T_{h=1}(-k,k)] + \cdots.\end{aligned} \tag{3.78}$$

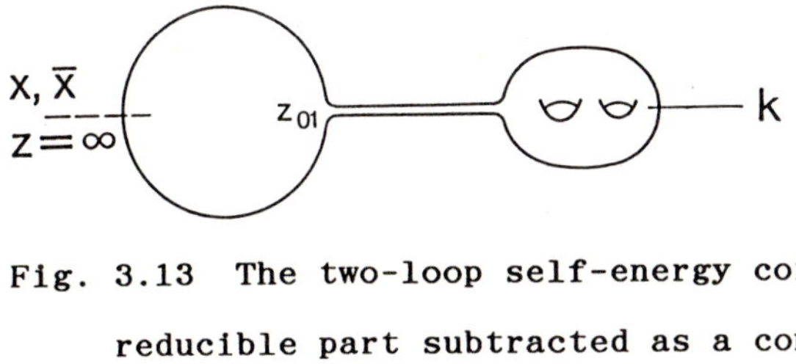

Fig. 3.13 The two-loop self-energy correction with the one-particle reducible part subtracted as a conformal symmetry breaking at z_{01}.

When we shift the anomalous dimension by the self-energy corrections, we obtain a modified W-function

$$\tau(x) = 1+\int dk\int d^2z\varepsilon^{k^2-2-[T(-k,k)]_{1PI}} e^{ik\mathfrak{X}(z)} W(\phi,k). \tag{3.79}$$

It is important to note that we have found the product of the one-loop self-energy and the one-loop vertex terms in eq.(3.65) which is precisely needed to make the modified W-function free from the self-energy corrections. Therefore we can summarize the result up to the two-loop order as the shift of the anomalous dimension (3.78) and the remaining W-function free from self-energy contributions. We remark that the tadpole contribution (N=0) at two-loop order is not explicitly separated here.

3.4 Multi-loop case

We can perform an analogous analysis for the nonlinear sigma model on genus h Riemann surfaces. We shall describe only self-energy corrections in some detail. The N-th order term in the weak field expansion is given by

$$\tau_h^N(x) = \frac{1}{(4\pi(8\pi^2)^{13})^h}\left(\frac{\kappa}{\pi}\right)^{2h}\int\prod_{s=1}^{h}\frac{d^2q_s}{|q_s|^4}\frac{1}{(\det \mathrm{Im}\Omega)^{13}}$$

$$\prod_i{}'\prod_{n\geq 1}|1-w_i{}^n|^{-48}\Gamma\int\prod_{i=1}^{N}dk_i\phi(k_i)\bar{\tau}_h^N \tag{3.80}$$

$$\bar{\tau}_h^N(x) = \frac{1}{N!}\int\prod_{i=1}^{N}d^2z_i\ e^{k_i{}^2-2}\int\prod_{s=1}^{h}\frac{d^2z_{0s}d^2z_{\infty s}}{|z_{0s}-z_{\infty s}|^4}\ e^{i\sum_{i=1}^{N}k_i\mathfrak{X}(z_i)}$$

$$\prod_{1\leq i<j\leq N}(|z_i-z_j|^2+\varepsilon^2)^{k_ik_j}\exp\Big(-\sum_{1\leq r<\ell\leq N+1}\mathcal{G}k_rk_\ell\ (z_r,z_\ell;z_{01},z_{\infty 1};\cdots,;z_{0h}z_{\infty h})\Big).$$

Introducing rescaled variables

$$z^{(1)} = \frac{z-z_{01}}{z_{\infty 1}-z_{01}},\ \ldots,\ z^{(h)} = \frac{z-z_{0h}}{z_{\infty h}-z_{0h}}$$

$$z^{(h+1)} = \frac{z-z_1}{z_2-z_1}, \tag{3.81}$$

we change variables to obtain

$$\bar{\tau}_h^N(x) = \frac{1}{N!}\int d^2 z_{01} \varepsilon^{\Sigma(k_i^2-2)} e^{i\sum_{i=1}^{N} k_i \mathfrak{X}(z_{01})}$$

$$\int d^2 z_\infty (|z_{\infty 1}-z_{01}|^2+\varepsilon^2)^{\Sigma k_i k_j+N-2} \int \prod_{s=2}^{h} d^2 z_{0s}^{(s-1)} d^2 z_{\infty s}^{(s-1)}$$

$$(|z_{\infty s}^{(s-1)}-z_{0s}^{(s-1)}|^2+\frac{\varepsilon^2}{|z_{\infty s-1}^{(s-2)}-z_{0s-1}^{(s-2)}|^2\cdots|z_{\infty 1}-z_{01}|^2})^{\Sigma k_i k_j+N-2} \times J+\cdots,$$

$$J = \int d^2 z_1^{(h)} d^2 z_2^{(h)} (|z_2^{(h)}-z_1^{(h)}|^2 \tag{3.82}$$

$$+\frac{\varepsilon^2}{|z_{\infty h}^{(h-1)}-z_{0h}^{(h-1)}|^2\cdots|z_{\infty 1}-z_{01}|^2})^{\Sigma k_i k_j+N-2}$$

$$\exp(-\sum_{1\leq r<\ell\leq N+1} k_r k_\ell \mathcal{G}(z_r,z_\ell;z_{01},z_{\infty 1};\cdots;z_{0h},z_{\infty h})).$$

A partial integration in $z_2^{(h)}$ gives two terms: a pole term multiplied by a dual N+1 point function at the tree level and a remaining term obtained as an analytic continuation from $\Sigma k_i k_j+N-2 \gtrsim -1$. Let us concentrate our attention to the most singular contribution obtained from successive partial integrations in $z_2^{(h)}$, $z_{\infty h}^{(h-1)}$, $\cdots$, $z_{\infty 1}$ variables

$$\tau_h^N|_{\text{most singular}} = \frac{1}{N!}\frac{-2\pi}{2(\Sigma k_i k_j+N-1)} t_0^{(N+1)}(-k,k_1,\cdots,k_N)\tau_h^{N=1}|_{\text{most singular}}$$

$$\bar{\tau}_h^{N=1}|_{\text{most singular}} = \int d^2 z_{01} \varepsilon^{\Sigma(k_i^2-2)} e^{ik\mathfrak{X}(z_{01})}$$

$$\cdot\int \prod_{s=1}^{h} d^2 z_{\infty s}^{(s-1)} d^2 z_{0s+1}^{(s)} (|z_{\infty s}^{(s-1)}-z_{0s}^{(s-1)}|^2+\frac{\varepsilon^2}{|z_{\infty s}^{(s-1)}-z_{0s}^{(s-1)}|^2\cdots|z_{\infty 1}-z_{01}|^2})^{-1}$$

$$\cdot\exp(k^2\mathcal{G}^{\tau_s}(z_{0s+1}^{(s)}))$$

$$\cdot\int d^2 z_{\infty h}^{(h-1)} d^2 z_1'' (|z_{\infty h}^{(h-1)}-z_{0h}^{(h-1)}|^2+\frac{\varepsilon^2}{|z_{\infty h-1}^{(h-2)}-z_{0h-1}^{(h-2)}|^2\cdots|z_{\infty 1}-z_{01}|^2})^{-1}$$

$$\cdot e^{k^2\mathcal{G}^{\tau_h}(z''_1)}. \tag{3.83}$$

The integral can be evaluated by using an identity

$$\int_\varepsilon^\infty \frac{dx_1}{x_1}\int_{\varepsilon/x_1}^\infty \frac{dx_2}{x_2}\cdots\int_{\varepsilon/x_1\cdots x_{h-1}}^\infty \frac{dx_h}{x_h} = \frac{1}{h!}\left(\ln\frac{1}{\varepsilon}\right)^h$$

which is easily proved by induction. Collecting the most singular part we obtain

$$\sum_{h=0}^{\infty} \tau_h^{N=1} \Big|_{\text{most singular}}$$

$$= \sum_{h=0}^{\infty} \int dk\phi(k) \int d^2z \; \varepsilon^{k^2-2} \; e^{ik\mathcal{X}(z)} \frac{1}{h!}\left(T_{h=1}(-k,k)\frac{1}{2}\ln\frac{1}{\varepsilon^2}\right)^h$$

$$= \int dk\phi(k) \int d^2z \; \varepsilon^{k^2-2-T_{h=1}(-k,k)} \; e^{ik\mathcal{X}(z)} . \qquad (3.84)$$

Thus we find explicitly that the one-loop self-energy contribution exponentiates and becomes a shift of the anomalous dimension.

References

[1] D. Friedan, Phys. Rev. Lett. 45(1980) 1057

[2] L. Alvarez-Gaumé, D. Freedman and S. Mukhi, Ann. of Phys. 134(1981)85

[3] C. Lovelace, Phys. Lett. B135(1984)75

[4] E. Fradkin and A. Tseytlin, Nucl. Phys. B261(1985)1

[5] P. Candelas, G. Horowitz, A. Strominger, and E. Witten, Nucl. Phys. B258(1985)46

[6] S. Jain, G. Mandal and S. Wadia, Phys. Rev. D32(1985)2713

[7] C. Callan, D. Friedan, E. Martinec and M. Perry, Nucl. Phys. B262(1985)593; A. Sen, Phys. Rev. D32(1985)2102

[8] B. Fridling and A. Jevicki, Phys. Lett. B174(1986)75

[9] T. Banks, D. Nemeshansky and A. Sen, Nucl. Phys. B277(1986)67

[10] S.R. Das and B. Sathiapalan, Phys. Rev. Lett. 56(1986)2664; 57(1986)1511

[11] C. Lovelace, Nucl. Phys. B273(1986)413

[12] R. Akhoury and Y. Okada, Phys. Lett. B183(1987)65

[13] Y. Watabiki, prep. TIT/HEP-114 (1987)

[14] I.G. Koh and H.J. Shin, prep. KAIST 87/2 (1987)

[15] G. Aldazabal, A. Diaz and R. Zhang, prep. IC/87/89; G. Keller and R. Silvotti, ETH prep.

[16] N. Sakai and Y. Tanii, Nucl. Phys. B287(1987)457

[17] M. Abe, H. Kubota and N. Sakai, prep. TIT/HEP-121, Phys. Lett. in press

[18] J. Ellis, P. Jetzer and L. Mizrachi, prep. CERN-TH.4829/87

[19] W. Fischler and L. Susskind, Phys. Lett. B171(1986)383; B173(1986)262

[20] C. Callan, C. Lovelace, C. Nappi and S. Yost, Nucl. Phys. B; Princeton prep. PUPT-1045

[21] S.R. Das and S.J. Rey, Phys. Lett. B186(1987)328; prep. CALT-68-1454

[22] J. Polchinski and Y. Cai, Texas prep. UTTG-09-87

[23] W. Fischler, I. Klebanov and L. Susskind, SLAC-Pub-4298

[24] H. Ooguri and N. Sakai, Phys. Lett. B197(1987)109

[25] J.M. Cline, prep. CALT-68-1458; A. Sen, prep. SLAC-Pub-4383;R. Metsaev and A. Tseytlin, prep. Lebedev N193

[26] V. Alessandrini, Nuovo Cimento 2A(1971)321; C. Lovelace, Phys. Lett. 32B(1970)703; V. Alessandrini and D. Amati, Nuovo Cimento 4A(1971)793; M. Kaku and L.P. Yu, Phys. Rev. D3(1971)2992, 3007, 3020; M. Kaku and J. Scherk, Phys. Rev. D3(1971)430, 2000

[27] S. Mandelstam, in Unified String Theories, Proc. Santa Barabara Workshop, p.46 (World Scientific, Singapore, 1986)

[28] P. Divecchia, M. Frau, A. Lerda, and S. Sciuto, preprint NORDITA-87/50 P

[29] L. Ford, Automorphic Functions (Chelsea, New York, 1951)

[30] N. Ishibashi, Y. Matsuo and H. Ooguri, Mod. Phys. Lett. 2(1987)119; prep. UT-499

[31] L. Alvarez-Gaumé, C. Gomez and C. Reina, Phys. Lett. B190(1987)55

[32] C. Vafa, Phys. Lett. B190(1987)47; K. Narain and T. Jayaraman, Rutherford prep. RAL-87-111

[33] N. Kawamoto, Y. Namikawa, A. Tsuchiya and Y. Yamada, prep. Nagoya Univ. Dept. Math.

[34] A.M. Polyakov, SLAC-TRANS-0222(1986)

[35] A.B. Zamolodchikov, JETP Lett. 43(1986)730

[36] T. Banks and E. Martinec, Nucl. Phys. B294(1987)733

[37] B. Sathiapalan, Nucl. Phys. B294(1987)747

[38] R. Brustein, D. Nemeschansky and S. Yankielowicz, prep. USC-87/004

[39] R.I. Nepomechie, Phys. Rev. D32(1985)3201

[40] J. Scherk, Nucl. Phys. B31(1971)222; A. Neveu and J. Scherk, Nucl. Phys. B36(1972)155

[41] T. Yoneya, Nuovo Cim. Lett. 8(1973)951; Prog. Theor. Phys. 51(1974)1907; J. Scherk and J.H. Schwarz, Nucl. Phys. B81(1974)118

[42] E. D'Hoker and D.H. Phong, Phys. Rev. D35(1987)3890

[43] E. Date, M. Kashiwara, M. Jimbo and T. Miwa, in Proc. RIMS Symp. on Nonlinear Intergrable Systems (Kyoto, 1981) p.39, eds. M. Jimbo and T. Miwa (World Scientific, Singapore, 1983)

[44] D. Mumford, Tata Lectures on Theta I and II (Birkhäuser, Boston, 1984)

[45] G. Moore, Phys. Lett. B176(1986)369

[46] J. Fay, Theta Functions on Riemann Surfaces, Lecture Notes in Mathematics, Vol.352 (Springer, Berlin, 1973)

[47] H. Ooguri and N. Sakai, prep. TIT/HEP-123 UT-519

Differential Equations for Conformal Characters in Moduli Space

T. Eguchi

Department of Physics, University of Tokyo, Tokyo, Japan 113

It is shown that character functions of degenerate representations of Virasoro algebra satisfy a linear differential equation in moduli space. In the case of a torus, we obtain an ordinary differential equation in variable τ of order $m(m-1)/2$ (m is related to the center c as $c=1-6/m(m+1)$) with coefficient functions given by elliptic theta constants.

It is well-known that correlation functions of minimal conformal models obey a set of linear differential equations coming from the existence of null states.[1] Solutions of these equations are expressed in terms of integral representations and are used to determine operator-product expansion coefficients.[2,3]

In this note we discuss conformal field theory on Riemann surfaces and point out that partition functions of minimal models themselves obey differential equations of a similar nature.[4] In the case of a torus we obtain an ordinary differential equation in the variable τ of order $m(m-1)/2$ (m is related to the central charge of Virasoro algebra as $c=1-6/m(m+1)$) for conformal character functions. These differential equations follow from the existence of a null state of degree $m(m-1)$ in the module of highest-weight state $h=0$ corresponding to the identity operator.

Let us first review basic steps in deriving differential equations of correlation functions. In the case of a complex plane, conformal Ward identity is given by[1,5]

$$\langle T(z)\phi_i(w_i)\phi_2(w)_2 \cdots\rangle$$

$$= \sum_{i=1} \{\frac{h_i}{(z-w_i)^2} + \frac{1}{z-w_i}\partial_{w_i}\}\langle \phi_1(w_1)\phi_2(w)_2 \cdots\rangle \quad . \tag{1}$$

Here ϕ_i's are conformal fields and h_i's are their conformal dimensions. $T(z)$ is the holomorphic part of the energy-momentum tensor. For the sake of illustration let us now consider the case of the Ising model which corresponds to the $m=3$, $c=1/2$ representation of Virasoro algebra. It is known that in this case there exists, for instance, a degeneracy at the 2nd grade in the module of highest-weight state $h=1/2$

$$(L_{-2} - \frac{3}{4}(L_{-1})^2)|h = 1/2\rangle = 0 \tag{2}$$

and this state must decouple from the rest of states in a unitary theory. Making use of the operator product expansion

$$T(z)\phi(w) \approx \{\frac{h}{(z-w)^2} + \frac{1}{z-w}\partial_w\}\phi(w) + \cdots\rangle$$

$$(T(z) = \sum_n L_n(w)(z-w)^{-n-2}) \tag{3}$$

Springer Proceedings in Physics Vol. 31: **Quantum String Theory**
Editors: N. Kawamoto and T. Kugo

and using the fact that $L_{-1}(w)\phi(w) = \partial_w \phi(w)$ we obtain

$$\frac{3}{4}\frac{\partial^2}{\partial w_1^2}\langle\phi_1(w_1)\phi_2(w_2)\cdots\rangle$$

$$= \sum_{i=2}\{\frac{h_i}{(w_1-w_i)^2} + \frac{1}{w_1-w_i}\partial_{w_i}\}\langle\phi_1(w_1)\phi_2(w_2)\ \cdots\rangle \tag{4}$$

when ϕ_1 is a conformal field with $h=1/2$. (4) is a prototype of the differential equations in conformal field theory. To each null state in representation space corresponds a linear differential operator whose order is given by the grade of the null state.

In the case of a Riemann surface the Ward identity acquires a new term which describes the dependence of correlation functions on the moduli of the Riemann surface. If we consider the case of a torus for simplicity, the conformal Ward identity is given by[6]

$$\langle T(z)\phi_1(w_1)\phi_2(w_2)\ \cdots\rangle$$

$$= \sum_i\{h_i(\wp(z-w_i) + 2\eta_1) + (\zeta(z-w_i) + 2\eta_1 w_i)\partial_{w_i}\}\langle\phi_1(w_i)\phi_2(w_2)\ \cdots\rangle$$

$$+ \frac{2\pi i}{Z(\tau)}\ \frac{\partial}{\partial\tau}\{Z(\tau)\langle\phi_1(w_1)\phi_2(w_2)\ \cdots\rangle\}\ . \tag{5}$$

Here $Z(\tau)$ is the partition function and $\wp$, ζ are Weierstrass $\wp$ and ζ functions. $\eta_1 = \zeta(1/2) = 2\pi i\partial_\tau \log(1/\eta(\tau))$ where $\eta(\tau)$ is the Dedkind η-function. Local algebraic structure remains the same on the Riemann surface and we can still use the equation of degeneracy (2). Then following the same steps leading to (4) we obtain

$$\{\frac{3}{4}\partial^2_{w_1} - \eta_1 - 2\eta_1 w_1\partial_{w_1}\}\langle\phi_1(w_1)\phi_2(w_2)\ \cdots\rangle$$

$$= \sum_{i=2}\{h_i(\wp(w_1-w_i)+2\eta_1) + (\zeta(w_1-w_i)+2\eta_1 w_i)\partial_{w_i}\}\langle\phi_1(w_1)\phi_2(w_2)\ \cdots\rangle$$

$$+ \frac{2\pi i}{Z(\tau)}\frac{\partial}{\partial\tau}\{Z(\tau)\langle\phi_1(w_1)\phi_2(w_2)\ \cdots\rangle\}\quad . \tag{6}$$

Differential equations of the type (6) have been used to calculate Ising correlation functions on the torus.[7]

Now let us discuss differential equations on conformal character functions. We stick to the case of the Ising model and consider null states generated out of the identity operator $h=0$. From the Kac formula[8] we know that there exists a null state at grades 1 and 6,

$$L_{-1}(z)\mathbb{1} = 0$$

$$\{64(L_{-2}(z))^3+93(L_{-3}(z))^2-264L_{-4}(z)-108L_{-6}(z)\}\ \mathbb{1} \equiv \Phi(z) = 0\quad . \tag{7}$$

The null state at grade 1 leads to a triviality, however, the one at grade 6 gives non-trivial relations. If we consider the case of a complex plane and compute, for instance, a 3-point function involving $\Phi(z)$, using the Ward identity (1) we find

$$\langle\Phi(z)\phi(w_1)\phi(w_2)\rangle = h(h - \frac{1}{2})(h - \frac{1}{16}) \frac{\text{const.}}{(z-w_1)^6(z-w_2)^6(w_1-w_2)^{2h-6}} , \tag{8}$$

where ϕ is a conformal field with dimension h. Thus the null state $\Phi(z)$ behaves as a conformal field and has vanishing operator-product expansion coefficients with other fields with dimension h=0, 1/2, 1/16. These are the allowed values of highest weights in the Ising model.

Now let us go to the case of a torus and consider a 1-point function of $\Phi(z)$. We obtain a differential equation of the partition function from the relation

$$\mathrm{tr}(\Phi(z)) = 0. \tag{9}$$

Here tr is taken over any of the highest weight representation $V_{h=0}$, $V_{h=1/2}$ $V_{h=1/16}$ of the Ising model. Making use of the Ward identity[6] and after some algebra we find

$$(2\pi i)^3 \frac{d^3}{d\tau^3} Z(\tau) + 12\eta_1(\tau)(2\pi i)^2 \frac{d^2}{d\tau^2} Z(\tau)$$

$$+ 2\pi i(- \frac{25}{64} g_2(\tau) + 24\, \eta_1(\tau)^2)\frac{d}{d\tau} Z(\tau) - \frac{23}{4\times 64} g^3(\tau)\, Z(\tau) = 0 , \tag{10}$$

where g_2 and g_3 are coefficients in the expansion of $\wp$-function

$$\wp(z) = \frac{1}{z^2} + \frac{1}{20} g_2\, z^2 + \frac{1}{28} g_3\, z^4 + \cdots . \tag{11}$$

(10) must be satisfied by each of the character functions $\chi_{h=0}$, $\chi_{h=1/2}$, $\chi_{h=1/16}$ of the representation $V_{h=0}$, $V_{h=1/2}$, $V_{h=1/16}$.

We make a change of variable from τ to λ given by

$$\lambda(\tau) = \frac{e_2(\tau) - e_3(\tau)}{e_1(\tau) - e_3(\tau)} , \tag{12}$$

where e_1, e_2, e_3 are related to g_2, g_3 as $g_2 = 2(e_1^2 + e_2^2 + e_3^2)$, $g_3=4e_1e_2e_3$ and obey $e_1+e_2+e_3=0$. λ is related to the parameters defining the torus as an elliptic curve, $y^2= x(x-x_0)(x-x_1)$, as $\lambda=x_0/x_1$. Then (10) can be rewritten as

$$\{\frac{d^3}{d\lambda^3} - \frac{2(2\lambda-1)}{(1-\lambda)\lambda} \frac{d^2}{d\lambda^2} + \frac{391(\lambda^2-\lambda)+7}{8\times 24(1-\lambda)^2\lambda^2} \frac{d}{d\lambda} - \frac{23}{24^3} \frac{(2-\lambda)(1+\lambda)(2\lambda-1)}{(1-\lambda)^3\lambda^3}\} Z(\lambda) = 0 . \tag{13}$$

In this representation (13) is a differential equation of Fuchsian type; coefficient functions are rational functions of λ, and λ=0, 1, ∞ are regular singular points. Modular transformation in the τ variable corresponds to monodromy transformation in λ variable. Solutions of (13) take particulary simple forms in terms of λ and in fact reproduce three character functions of the Ising Model,

$$\chi_{h=0} = \lambda^{-1/24}(1-\lambda)^{-1/24} + \lambda^{-1/24}(1-\lambda)^{1/12}$$

$$= \text{const.}\ q^{-1/48}\,\frac{1}{2}\{\prod_{n=1}(1+q^{n-1/2}) + \prod_{n=1}(1-q^{n-1/2})\}$$

$$\chi_{h=1/2} = \lambda^{-1/24}(1-\lambda)^{-1/24} - \lambda^{-1/24}(1-\lambda)^{1/12}$$

$$= \text{const.}\ q^{-1/48}\,\frac{1}{2}\{\prod_{n=1}(1+q^{n-1/2}) - \prod_{n=1}(1-q^{n-1/2})\}$$

$$\chi_{h=1/16} = \lambda^{-1/12}(1-\lambda)^{-1/24}$$

$$= \text{const.}\ q^{1/24}\prod_{n=1}(1+q^{n}) \quad . \tag{14}$$

In the general case null states occur at grade m(m-1) and the order of differential equation becomes m(m-1)/2 since each power of L_{-2} is converted into a derivative $d/d\tau$ by the Ward identity (note that L_{-1} does not occur since it annihilates the h=0 state). Then m(m-1)/2 independent solutions of the differential equation correspond to each of $\chi_{h_{r,s}}$ with $1\leq s\leq r\leq m-1$.

Differential equations for affine character functions can be derived in a similar manner.[9]

We thank Prof. K. Okamoto for a useful discussion on differential equations of Fuchsian type.

References

1. A. A. Belavin, A. M. Polyakov and A. B. Zamolodchikov, Nucl. Phys. B241 (1984) 333.
2. B. L. Feigin and D. B. Fucks, Moscow preprint 1983.
3. Vl. S. Dotsenko and V. Fatteev, Nucl. Phys. B240 (1984) 312.
4. T. Eguchi and H. Ooguri, preprint UT-518, Nov. 1988.
5. D. Friedan, in 1982 Les Houches summer school, Recent Advances in Field Theory and Statistical Mechanics, J. B. Suber and R. Stora (eds.), North-Holland (1984).
6. T. Eguchi and H. Ooguri, Nucl. Phys. B282 (1987) 308.
7. P. Di Francesco, H. Saleur and J. B. Zuber, Saclay preprints 1987.
8. V. Kac, Proceedings of the International Congress of Mathematics, Helsinki, 1978 and Lecture Notes in Physics 94 (1979) 441.
9. T. Eguchi and H. Ooguri, in preparation.

Part IV

Further Mathematical Developments

Analytic Fields on Riemann Surfaces

V.G. Knizhnik

Landau Institute, Moscow, USSR

It is known that string theories are often described by two-dimensional (2D) free quantum field theories on the world sheet. The specific feature of two dimensions is that the excitations of these fields are chiral and the left and right moving modes don't interact (locally). Thus it seems to be possible to reformulate the theory to describe these sectors directly in terms of QFT. This problem has 4 equivalent formulations:

1) To describe chiral sectors alone in terms of QFT.

2) To find the rules for the transformation to the 1st order formalism.

3) To find the way to compute the holomorphic roots $\det \overline{\partial}_j$ of $\det \Delta_j$, where Δ_j is the Laplace operator on a Riemann surface acting on j-differentials.

4) To pass from real to complex analytic language in 2D conformal field theory (CFT).

The "practical" purposes we have in mind are the following:

1) To compute string measure for various string theories (for example in the 26D bosonic string, the measure is expressed through $\det \overline{\partial}_{-1}(\det \overline{\partial}_0)^{-13} \equiv F_{BS}$).

2) To understand the specific features of CFT on a Riemann surface.

3) To construct new CFT.

Thus we start from a Riemann surface X with coordinates $\xi^{1,2}$, metric $g_{ab}(\xi)$ to which the conformal coordinates $z, \overline{z}$ correspond:

$$g_{ab}(\xi)d\xi^a d\xi^b = \rho(z,\overline{z})dz d\overline{z} \ . \tag{1}$$

We introduce the j-differential field $\phi_j(z,\overline{z})$ defined by the following transformation law:

$$\phi_j(z,\overline{z}) = \left(\frac{df}{dz}\right)^j \widetilde{\phi}_j(f,\overline{f}) \tag{2}$$

and consider only $j \in \frac{1}{2}\mathbf{Z}$. The action for such a field has the form

Springer Proceedings in Physics Vol. 31: **Quantum String Theory**
Editors: N. Kawamoto and T. Kugo

$$S^{(j)}_{\mathrm{Bos}} = \int \rho^{-j} |\overline{\partial}\phi_j|^2 d^2 z \ , \tag{3}$$

where $\overline{\partial} \equiv \frac{\partial}{\partial \overline{z}}$, $d^2 z \equiv i dz \wedge d\overline{z}$. To proceed to the 1st order formulation we need a pair of *anticommuting* fermionic fields $\phi^{(j)}$ and $\phi^{(1-j)}$ to write the covariant action

$$S^{(j)}_F = \int \phi^{(1-j)} \overline{\partial} \phi^{(j)} d^2 z \ , \tag{4}$$

and our problem is to find the correct definition for the partition function for (4). One can try simply

$$\int \mathcal{D}\phi^{(j)} \mathcal{D}\phi^{(1-j)} \exp[-S^{(j)}_F] \ ,$$

but it turns out to be zero due to zero modes of $\phi^{(j)}$, $\phi^{(1-j)}$-fields. To extract the zero mode contribution we have to choose for each j a basis

$$\left\{ \phi^{(j)}_\alpha(z) \ , \quad \alpha = 1 \cdots n_j \right\}$$

in the space of holomorphic j-differentials and write

$$\det \overline{\partial}_j = \frac{\left\langle \phi^{(j)}(z_1) \cdots \phi^{(j)}(z_{n_j}) \phi^{(1-j)}(z'_1) \cdots \phi^{(1-j)}(z'_{n_{1-j}}) \right\rangle}{\det \|\phi^{(j)}_\alpha(z_\beta)\| \det \|\phi^{(1-j)}_\gamma(z'_\delta)\|} \ , \tag{5}$$

where index j of $\overline{\partial}$ means that it acts on j-differentials, the $n_j \times n_j$ matrix $\|\phi^{(j)}_\alpha(z_\beta)\|$ has $\phi^{(j)}_\alpha(z_\beta)$ on the intersection of the raw α and the column β and

$$\langle \ \sqcup \ \rangle \equiv \int \ \sqcup \ e^{-S^{(j)}_F} \mathcal{D}\phi^{(j)} \mathcal{D}\phi^{(1-j)} \ . \tag{6}$$

We recall that for $p \gg 2$

$$\begin{aligned} & n_j = (2j-1)(p-1) \ , \quad j > 1 \\ & n_1 = p \\ & n_0 = 1 \\ & n_{1/2}(m) = e(m) \qquad \text{(in general)} \end{aligned} \tag{7}$$

and for $p = 1$

$$n_j = 1 \ , \quad j \in \mathbf{Z}$$

$$n_j(m) = e(m) \ , \quad j \in \mathbf{Z} + 1/2 \qquad \text{(in general)} \ ,$$

where characteristic m parametrises the boundary conditions on half-integer differentials (see appendix).

Actually, we expect some difficulty in the definition of (6) because chiral theories have the gravitational anomaly. Under general coordinate transformation

$$z \to z + \epsilon(z) ,$$

partition function has to change:

$$\begin{aligned} \delta_\epsilon \ln \det \overline{\partial}_j &= -\frac{C_j}{24\pi} \int \rho^{-1} \partial(\rho\epsilon)(\Delta \ln \rho) d^2 z \\ C_j &= 6j^2 - 6j + 1 . \end{aligned} \tag{8}$$

This relation has the following interpretation. In CFT with central charge C a point ξ_0 on a Riemann surface where the curvature concentrates according to

$$R(\xi) = 2\pi h \delta(\xi - \xi_0) \tag{9}$$

looks like a conformal field of dimension

$$\Delta_{\xi_0} = \frac{Ch}{12} . \tag{10}$$

For $S_F^{(j)}$ we have $C = -2C_j$ and thus

$$\Delta_{\xi_0} = -\frac{hC_j}{6} . \tag{11}$$

Thus we see that to solve (8) it is useful to choose a singular metric on X such that

$$R = \sum_i 2\pi h_i \delta(\xi - \xi_i) . \tag{12}$$

Below we discuss two choices of such metric and see the connection between theories of theta-functions of deformation equations and of twist fields.

Choice I

We fix the odd characteristic $*$ on a surface X and the corresponding holomorphic 1/2-differential

$$\nu_*(z)(dz)^{1/2} .$$

Let

$$\rho = |\nu_*|^4 . \tag{13}$$

It is known that on a genus p surface ν_* has exactly $p-1$ zeros, different in general. Let's denote them by $R_1 \cdots R_{p-1}$. Then

$$
\begin{aligned}
R &= \sum_i 8\pi\delta(\xi - R_i) \\
h(R_i) &= 4 \\
\Delta(R_i) &= -\frac{2}{3}C_j \; .
\end{aligned}
\tag{14}
$$

In order to guess the general formula for $\det \overline{\partial}_j$ we first consider:

a) The case $j = 0$

For $j = 0$ we can always choose the normalized basis ω_i, $i = 1 \cdots p$ of holomorphic 1-differentials (see appendix) and put the unique scalar zero mode

$$
\phi_1^{(0)}(z) = 1 \; . \tag{15}
$$

Then

$$
\det \overline{\partial}_0 = \frac{\left\langle \phi^{(1)}(z_1) \cdots \phi^{(1)}(z_p)\phi^{(0)}(z) \right\rangle}{\det \|\omega_i(z_j)\|} \tag{16}
$$

has to satisfy the following properties

(α) does not depend on z, z_i.

(β) is the modular form of weight 1 (with respect to transformations, preserving the characteristic $*$).

(γ) $\Delta(R_i) = -2/3$.

Surprisingly, it is easy to satisfy all these requirements by taking

$$
(\det \overline{\partial}_0)^{3/2} = \theta_{*,i}\omega_i(Q) \det{}^{-1} \|\vec{\omega}(Q)\vec{\omega}(R_1) \cdots \vec{\omega}(R_{p-1})\| \, , \tag{17}
$$

where $\theta_{*,i}$ is defined in appendix, Q is an arbitrary point and the matrix $\|\vec{\omega}(Q) \cdots \vec{\omega}(R_{p-1})\|$ is made of columns $\vec{\omega}(Q) = (\omega_i(Q) \cdots \omega_p(Q))^t$. One can prove that (17) does not depend on Q. This expression looks like a straightforward generalization of the $p = 1$ result

$$
(\det \overline{\partial}_0)^{3/2} = \theta'_{1,1} \qquad (\rho = 1) \; .
$$

The only tricky point is that it now depends on the choice of $*$ and of coordinates of points R_i, but this, as we have shown, can be interpreted as a consequence of the gravitational anomaly (8), (11), (14) for the special choice of a metric (13).

b) The case $j = 1/2$ (free fermions)

Here it is useful to change the notations and put

$$S_{1/2,m} = \int \psi \bar{\partial} \tilde{\psi} d^2 z \ ,$$

where ψ and $\tilde{\psi}$ are different 1/2-differentials satisfying boundary conditions m. Actually for this case we know that

$$\det{}_m \bar{\partial}_{1/2} (\det \bar{\partial}_0)^{1/2} = \theta_m(\tau) \ . \tag{18}$$

Let's introduce

$$v_*(z) = \sqrt{\theta_{*,i} \omega_i(z)} \tag{19}$$

(it can be shown that the root exists and $v_*/\nu_* =$ const.). One can show that we can uniquely reconstruct the correlation functions of $\psi, \tilde{\psi}$-fields from the boundary conditions and analytical properties (position of poles, residues, etc.) and get

$$\begin{aligned}
&\left\langle \psi(z_1) \cdots \psi(z_n) \tilde{\psi}(y_1) \cdots \tilde{\psi}(y_n) \right\rangle_m \\
&\equiv \int \psi(z_1) \cdots \tilde{\psi}(y_n) \exp[-S_{1/2,m}] \mathcal{D}\psi \mathcal{D}\tilde{\psi} \\
&= \det{}^{-1/2} (\bar{\partial}_0) \theta_m \Big(\sum_i (\vec{z}_i - \vec{y}_i)\Big) \\
&\times \prod_i v_*(z_i) v_*(y_i) \prod_{i<j} \theta_*(\vec{z}_i - \vec{z}_j) \theta_*(\vec{y}_i - \vec{y}_j) \\
&\times \prod_{i,j} \theta_*^{-1} (\vec{z}_i - \vec{y}_j) ,
\end{aligned} \tag{20}$$

where $\vec{z}_i - \vec{z}_j = \int_{z_j}^{z_i} \vec{\omega}(z) dz$, etc. For even m, and $n = 0$ we have (18), but for odd m there are fermion zero modes, $\theta_m \equiv 0$ and the new expression arises:

$$\det{}_m \bar{\partial}_{1/2} \det{}^{1/2} \bar{\partial}_0 = (v_m/\nu_m)^2 \ , \qquad e(m) = 1 \ , \tag{21}$$

where we have fixed the basis ν_* in 1 dimensional space of zero modes of $\bar{\partial}_{1/2,m}$ and substitute it and (20) into the general expression (5). With this examples at hand it is not difficult to know how to treat the general case.

c) The general case

It is useful to define not $\det \bar{\partial}_j$ but

$$\lambda_j \equiv \det \bar{\partial}_j \det{}^{1/2} \bar{\partial}_0 \ , \qquad j \gg 1 \ . \tag{22}$$

For this we have

$$\Delta(R_i) = -(2j-1)^2 \tag{23}$$

and it is convenient to describe the result in terms of correlation functions (20):

$$\begin{aligned}\left\langle \phi^{(j)}(z_1)\cdots\phi^{(1-j)}(z'_{n_{1-j}})\right\rangle \det{}^{1/2}\overline{\partial}_0 \\ = \left\langle \prod_{i=1}^{n_j} v_*^{2j-1}(z_i)\psi(z_i) \prod_{k=1}^{n_{1-j}} v_*^{1-2j}(z_k)\widetilde{\psi}(z_k) \prod_{\ell=1}^{p-1} V_j(R_\ell)\right\rangle_m ,\end{aligned} \tag{24}$$

where

$$V_j(R) \equiv : \widetilde{\psi}\partial\widetilde{\psi}\partial^2\widetilde{\psi}\cdots\partial^{2j-2}\widetilde{\psi}(R) : \left(\partial v_*(R)\right)^{-(2j-1)^2} \tag{25}$$

and

$$\begin{aligned} &m = * \qquad \text{for} \qquad j \in \mathbf{Z} \\ &m = \text{characteristic of } \phi^{(j)},\ \phi^{(1-j)} \qquad \text{for} \qquad j \in 1/2 + \mathbf{Z} .\end{aligned} \tag{26}$$

We note that the rule

$$\begin{aligned} \phi^{(j)}(z) &\longrightarrow v_*^{2j-1}(z)\psi(z) \\ \phi^{(1-j)}(z) &\longrightarrow v_*^{1-2j}(z)\widetilde{\psi}(z) ,\end{aligned} \tag{27}$$

which can be read off from (24), is also valid for other correlation functions of fields $\phi^{(j)}$, $\phi^{(1-j)}$.

d) Verification

1) $\lambda_0 =$ (previous expression (17))$^{3/2}$
   To prove it one has to take the limit $z_1 \to R_1, \cdots, z_{p-1} \to R_{p-1}, z_p \to Q$ in (16).

2) λ_j does not depend on z_i, z'_k and has modular weight 1/2.

3) $\Delta(R_i) = -(2j-1)^2$

Now we describe some particular cases interesting for string theory.

e) Fermionic ghost determinant: $j = 3/2$

Taking the limit $z_i, z_{p-1+i} \to R_i$, $i = 1\cdots p-1$ in (5), (24) for $j = 3/2$ we obtain for *even* m

$$\lambda_{3/2}(m) = \theta_m \det{}^{-1} \| \vec{\xi}(R_1)\partial\vec{\xi}(R_1)\cdots\vec{\xi}(R_{p-1})\partial\vec{\xi}(R_{p-1})\| , \tag{28}$$

where $\vec{\xi}(R) \equiv (\xi_1(R)\cdots\xi_{2p-2}(R))^t$ – the column of linearly independent 3/2 differentials ξ_i at a point R, $\partial\vec{\xi}(R) \equiv (\partial\xi_1(R_1)\cdots\partial\xi_{2p-2}(R_1))^t$.

f) Bosonic string measure F

$$
\begin{aligned}
F =& \lambda_2 \lambda_0^{-g} \\
=& [\partial^2 v_*(R_1)]^{-8} W^{-1} \det{}^9 \| \vec{\omega}(R_1) \partial^2 \vec{\omega}(R_1) \vec{\omega}(R_2) \cdots \vec{\omega}(R_{p-1}) \| \, .
\end{aligned} \tag{29}
$$

We used that

$$
\begin{aligned}
\lambda_2 &= (\partial^2 v_*) \cdot W^{-1} \\
W &\equiv \det \| \vec{f}(R_1) \partial^2 \vec{f}(R_1) \partial^4 \vec{f}(R_1) \vec{f}(R_2) \cdots \partial \vec{f}(R_{p-1}) \partial^2 \vec{f}(R_{p-1}) \| \, ,
\end{aligned} \tag{30}
$$

where $\vec{f} = (f_1 \cdots f_{3p-3})^t$ is a column of holomorphic 3-differentials used to define coordinates y_i on the moduli space $\mathcal{M}_p$ of genus p Riemann surfaces, partition function of which is given by

$$
Z_p = \int_{\mathcal{M}_p} \prod_{i=1}^{3p-3} dy_i \wedge d\overline{y}_i |F(y)|^2 (\det Im\tau)^{-13} \, . \tag{31}
$$

Again (30) is obtained by taking the limit $z_i, \; z_{p-1+i}, \; z_{2p-2+i} \to R_i \, , \; i = 1 \cdots p-1$ in (24) for λ_2.

Choice II

Here the metric ρ is chosen to be lifted from $\boldsymbol{CP}^1$ by a ramified covering

$$
\begin{aligned}
& y : X \xrightarrow{y} \boldsymbol{CP}^1 \\
& \rho_y(z, \overline{z}) dz d\overline{z} = dy d\overline{y} \, .
\end{aligned} \tag{32}
$$

The curvature is now concentrated at infinity and the ramification points. It can be shown in general that a simple ramification point of the order n in the theory with central charge C has the conformal dimension

$$
\Delta_n = \frac{C}{24} (n - \frac{1}{n}) \tag{33}
$$

with respect to y-transformations. One can choose the coordinate z or X near the ramification point a in such a way that

$$
y = a + z^n \, . \tag{34}
$$

Let's numerate the successive sheets of the inverse map

$$
z = (y - a)^{1/n}
$$

by numbers $0, 1, \cdots n-1$ and on each sheet we introduce a pair of analytic fields ${}^{\ell}\phi^{(j)}$, ${}^{\ell}\phi^{(1-j)}$, the action being

$$S^{(j)} = \sum_{\ell} S^{(j)}_{\ell}\,, \qquad S^{(j)}_{\ell} = \int_{\text{sheet } \ell} {}^{\ell}\phi^{(j)}\bar{\partial}\ {}^{\ell}\phi^{(1-j)} d^2y\,. \tag{35}$$

These fields satisfy the following boundary condition:

$$\widehat{\pi}_a\ {}^{\ell}\phi^{(\cdot)} = {}^{\ell+1}\phi^{(\cdot)}, \tag{36}$$

where $n-1+1 \equiv 0$ and $\widehat{\pi}_a$ denotes the operation of analytic continuation around the ramification point a. As for the Choice I, the conformal field $V_n(a)$ corresponds to the ramification point.

Remarkably, $V_n(a)$ can be expressed explicitly in terms of ${}^{\ell}\phi^{(\cdot)}$ fields! To do this we introduce the $\widehat{\pi}_a$ eigenvalues

$$\phi^{(j)}_{k\,(1-j)} = \frac{1}{\sqrt{n}} \sum_{\ell} {}^{\ell}\phi^{(j)}_{(1-j)} \exp[\mp \frac{2\pi i}{n} k(\ell + j(1-n))] \tag{37}$$

and bosonize them using n bosonic fields φ_k, $k = 1 \cdots n$:

$$\left\langle \varphi_k(y)\varphi_{\ell}(y') \right\rangle = -\delta_{k,\ell} \ln(y - y')\,. \tag{38}$$

<u>Bosonization rules (proof omitted)</u>

$$\begin{aligned}
&\phi^{(j)}_k(y) =: \exp i\varphi_k(y) : \\
&\phi^{(1-j)}_k(y) =: \exp(-i\varphi_k(y)) : \\
&V_n(a) =: \exp(\sum_{k=0}^{n-1} iq_k\varphi_k(a)) : \\
&q_k = \frac{k + j(1-n)}{n}\,.
\end{aligned} \tag{39}$$

The fields φ_k also have to have the charge $2j-1$ at $y = \infty$ thus*

$$V^{(\infty)} = \exp(2j-1)(\sum i\varphi_k(\infty)) \tag{40}$$

and the energy-momentum tensor of φ_k-field is

$$T_k = -\frac{1}{2}(\partial\varphi_k)^2 + \frac{i}{2}(1-2j)\partial^2\varphi_k\,. \tag{41}$$

The determinants now are expressed through the averages $\langle \prod_i V_{n_i}(a_i) \rangle$ and one can compute them using (39),(40). The nontrivial charge-conservation law for φ_k-field implied by (41) in

---

* Strictly speaking this has the definition only for the coverings with abelian monodromy group.

general enforce us to introduce into the average $\langle\prod_i V_{n_i}(a_i)\rangle$ the operators $\phi_k^{(j)}$, $\phi_\ell^{(1-j)}$ in a number needed to absorb all zero modes (this can be proved even for coverings with nonabelian monodromy group).

Examples

a) $\det\overline{\partial}_0$ *on a hyper-elliptic surface*

Let X be defined by the equation

$$z^2 = (y - a_1)\cdots(y - a_{2p+2}) \,. \tag{42}$$

For $j = 1$ we have from (39)

$$a_0 = -1/2\,, \qquad a_1 = 0$$

$$V_2(a_i) =: \exp(-\frac{i}{2}\varphi_0(a_i)) : \,.$$

From (40) follows that one has to insert p 1-differentials $\phi_0^{(1)} = e^{i\varphi_0}$ and 1 scalar $\phi_0^{(0)} = e^{-i\varphi_1}$. Thus we get

$$\begin{aligned}
&\Big\langle \phi_0^{(1)}(y_1)\cdots\phi_0^{(1)}(y_p)\phi_0^{(0)}(y)\prod_{i=1}^{2p+2} V_2(a_i)\Big\rangle \\
&= \prod_{i<j}(y_i - y_j)\prod_{i=1}^{p} z^{-1}(y_i)\prod_{k<\ell}^{2p+2}(a_k - a_\ell)^{1/4} \\
&= \det\overline{\partial}_0 \det\|\omega_i(y_k)\| \,.
\end{aligned} \tag{43}$$

Applying $\prod_{i=1}^{p}\oint_{a_i} dy_i$ we obtain

$$\det\overline{\partial}_0 = \det K \prod_{k<\ell}^{2p+2}(a_k - a_\ell)^{1/4} \,, \tag{44}$$

where

$$K_{i\ell} = \oint_{a_i} y^{\ell-1} z^{-1}(y)dy\,; \quad i,\ell = 1\cdots p\,.$$

From this in particular we obtain

$$\begin{aligned}
\det\Delta_0 &= |\det K|^2 \prod_{k<\ell}|a_k - a_\ell|^{1/2} \det Im\tau \\
&= d^2y_1\cdots d^2y_p\Big|\prod_{i<j}(y_i - y_j)\prod_i z^{-1}(y_i)\prod_{k<\ell}(a_k - a_\ell)^{1/4}\Big|^2 \,.
\end{aligned} \tag{45}$$

b) $j = 1/2$ on (42)

For $j \in \mathbf{Z} + 1/2$ we have sign ambiguity in the right hand side of (36)

$$\widehat{\pi}_a{}^{\ell} \phi^{(\cdot)} = \pm^{(\ell+1)} \phi^{(\cdot)} . \tag{46}$$

For X defined by (42) and $j = 1/2$ this leads to 2 possibilities

$$V_2^{\pm}(a_i) = \exp\{\mp\frac{i}{4}\varphi_0(a_i) \pm \frac{i}{4}\varphi_1(a_i)\} . \tag{47}$$

From this we see that characteristics m for which there are no fermion zero modes are described by the partitions of points a_i over two groups of $p+1$ points

$$\underset{2p+2}{\{a_i\}} = \underset{p+1}{\{a_i'\}} \cup \underset{p+1}{\{a_i''\}} \to m \equiv \{a_i'|a_i''\} . \tag{48}$$

For this we have

$$\begin{aligned} \det{}_m \bar{\partial}_{1/2} &= \left\langle \prod_i V_2^+(a_i')V_2^-(a_i'') \right\rangle \\ &= \prod_{i<j}(a_i' - a_j')^{1/8}(a_i'' - a_j'')^{1/8} \prod_{i<j}(a_i' - a_j'')^{-1/8} . \end{aligned} \tag{49}$$

Note that from (18), (44) and (49) one can obtain the Thomae formula for θ_m on a hyperelliptic curve.

Nonabelian monodromy group

This case is in close relations with papers by JMS= Jimbo, Miwa, Sato (RIMS, Kyoto) on the solution of Riemann problem: to find linear differential equation with given monodromy group

Let $X \underset{y}{\longrightarrow} \boldsymbol{CP}^1$ be N-sheeted covering with ℓ simple ramification points a_i of orders n_i. Operators $V_i(a_i)$ corresponds to them and suppose that charges are adjusted in such a way that

$$\tau(a_1 \cdots a_\ell) \overset{\text{def}}{=} \left\langle \prod_{i=1}^{\ell} V_{n_i}(a_i) \right\rangle \neq 0 . \tag{50}$$

Let us change the notations

$$\phi^{(j)} \equiv f , \qquad \phi^{(1-j)} \equiv \phi .$$

We consider the Green function

$$Y^{km}(y) = (y_0 - y)\frac{\left\langle \phi^k(y_0) f(y) \prod_i V_{n_i}(a_i) \right\rangle}{\tau} , \tag{51}$$

where $\phi^k(f^k)$- are the fields $\phi(f)$ taken on the sheet k. We have

$$\hat{\pi}_{a_i} Y = Y \cdot M_i \,, \tag{52}$$

where the matrices M_i represent permutations (for $j \in \mathbf{Z}$) and thus does not depend on y_0 or a_i. From this it follows that functions $\partial_y Y \cdot Y^{-1}$, $\partial_{a_i} Y \cdot Y^{-1}$, $\partial_{y_0} Y \cdot Y^{-1}$ are rational functions of z and from the OPE's of $\phi^k(f^k)$ with $V_{n_i}(a_i)$ and normalization condition

$$Y^{km}(y_0, y_0) = \delta^{km} \tag{53}$$

it follows that

$$\begin{aligned}
\frac{\partial Y}{\partial y} Y^{-1} &= \sum_{i=1}^{\ell} \frac{A_i}{y - a_i} \\
\frac{\partial Y}{\partial a_i} Y^{-1} &= \frac{A_i}{y_0 - a_i} - \frac{A_i}{y - a_i} \\
\frac{\partial Y}{\partial y_0} Y^{-1} &= -\sum_{i=1}^{\ell} \frac{A_i}{y_0 - a_i} \,.
\end{aligned} \tag{54}$$

From the conditions of the compatibility of these equations we get well-known Schlesinger deformation equations for A_i(JMS):

$$dA_m = \sum_k{}' [A_k, A_m] \, d \ln \left(\frac{a_m - a_k}{y_0 - a_k} \right) , \tag{55}$$

where

$$d \equiv dy_0 \frac{\partial}{\partial y_0} + \sum_i da_i \frac{\partial}{\partial a_i} \,. \tag{56}$$

Next we can insert into (51) the following OPE:

$$\begin{aligned}
\sum_k f^k(y) \phi^k(y_0) =& N(y - y_0)^{-1} + J(y_0) \\
&+ (y - y_0)[j \partial J(y_0) + T(y_0)] + O((y - y_0)^2) \\
J \equiv& \sum_k : f^k \phi^k : \,.
\end{aligned} \tag{57}$$

From this we find the value of the energy-momentum tensor in terms of A_i and their derivatives and compairing it with

$$\langle T(y_0) \rangle = \sum_i \left(\frac{A_{n_i}}{(y_0 - a_i)^2} + \frac{1}{y_0 - a_i} \frac{\partial \ln \tau}{\partial a_i} \right) \,, \tag{58}$$

we obtain the famous result of JMS:

$$d\ln\tau = \sum_{i<k} tr(A_i A_k)\, d\ln(a_i - a_k)\ . \tag{59}$$

Thus for our M_i JMS τ-function is equal to θ-function. It would be interesting to see this connection in detail or to generalize it to other M_i, but this work remains to be done.

Conclusion

I think it is clear, now, that the transition to the analytic fields in conformal QFT is equivalent to the transition in geometry from real to complex analytic coordinates. It provides us the connections with algebraic geometry and deformation theory and the theory of soliton equations. Thus a possibility opens to unify a lot of ideas into the complex geometry at quantum strings.

At the end I would like to express a deep gratitude to professors K. Nishijima, N. Kawamoto and to Tetsuya Onogi for the kind hospitality at RIFP during the '87 October meeting in Kyoto.

Appendix

See any book on Riemann surfaces.

Note

This talk is based on the material present in my papers in PL **180B** (1986) 247 and CMP **112** (1987) 567 where all the references can be found. I apologize for the impossibility to bring them here.

On the Connection Between Zamolodchikov's W-Algebras and Kac-Moody Algebras

A.A. Belavin

Landau Institute, Moskow, USSR

The two dimensional conformal field theory (CFT) is important both for the description of the second order phase transitions in two dimensional statistical models and for the string theory. Thus it seems to be extremely useful to obtain the full classification of such theories. The example of so-called minimal models shows that for their construction the Virasoro algebra and its special degenerate representations are of great importance. But the possible symmetries in CFT are not exhausted by the Virasoro algebra. The following generalizations of the conformal algebra are known: superconformal Nevew-Schwarz-Ramond algebra, the extension with the current algebra, parafermionic algebras by A.B. Zamolodchikov and V.A. Fateev and also parafermionic algebras by A.B. Zamolodchikov [1]. Both in all known models of CFT connected with these symmetries and in minimal models, local operators are classified according to the special degenerate representations of these algebras.

All these facts lead to the idea that it is necessary to find all possible generalizations of the Virasoro algebra, including the latter as a subalgebra. In connection with this it is also important to understand the geometrical meaning of already known algebras. The Goddard-Kent-Olive(GKO)-like constructions shed light on the algebras of the parafermionic type. The parafermionic fields are the generators of the algebra which consists of the elements of the enveloping algebra $\widehat{s\ell}_k(2)$, commuting with its Heizenberg subalgebra.

In this talk I want to describe a certain construction which, I think, leads to W-algebras by A.B. Zamolodchikov. First, I recall their definition. W-algebras are generated by local conserved currents of higher spin $s \gg 3$. The simplest case is the W-algebra generated by the Laurent components W_n of the current with spin $s = 3$, together with the L_n-components of the energy-momentum tensor. The commutation relations have the following form:

$$[L_n, W_m] = (2n - m)W_{n+m}$$
$$[W_n, W_m] = \alpha_{n,m}L_{n+m} + \beta_{n,m}\sum_k : L_k L_{n+m-k} : +\gamma_n \delta_{n+m,0} C \,, \tag{1}$$

where $\alpha_{n,m}, \beta_{n,m}, \gamma_n$ are some numbers depending on n and m, and C is the central charge of the Virasoro algebra. L_n and W_n are not the generators of any Lie algebra but rather of the *associative* algebra with polynomial defining relations. The statement I want to make is that the

Springer Proceedings in Physics Vol. 31: **Quantum String Theory**
Editors: N. Kawamoto and T. Kugo

W-algebras by Zamolodchikov are connected with Kac-Moody (KM) algebras. That is, to each KM algebra there is a corresponding W-algebra. For example to $\widehat{s\ell}(2)$ the Virasoro algebra itself (the simplest W-algebra) corresponds, and to $\widehat{s\ell}(3)$ the algebra defined above. The construction which connects a current algebra $\widehat{\mathcal{G}}$ with a W-algebra is described below:

Let $\mathcal{G}$ be a simple Lie algebra with the Cartan-Weyl generators $X^\alpha, M^\alpha, Y^\alpha$ where $\alpha = 1, \cdots, N$. Then the current algebra $\widehat{\mathcal{G}}$ corresponding to it possesses the generators $X_n^\alpha, M_n^\alpha, Y_n^\alpha$ and the central charge "K", which satisfy the obvious relations. X_n^α generate subalgebra $\widehat{\mathcal{N}} \subset \widehat{\mathcal{G}}$. Let $\mathcal{V}_{\widehat{\mathcal{G}}}$ be the enveloping algebra of $\widehat{\mathcal{G}}$ i.e. it consists of series in elements of $\widehat{\mathcal{G}}$ ordered in some definite way. Let also $\widehat{N}$ denotes the left ideal generated by the elements of the type $a - \chi(a)$ where $a \in \widehat{\mathcal{N}}$ and $\chi(a)$ is a nontrivial 1-dimensional representation (a character) of the algebra $\widehat{\mathcal{N}}$ (i.e. $\chi([a_1, a_2]) = 0$) such that

$$\chi(X_n^\alpha) = \delta_{n,0} \cdot$$

That is, we define $\widehat{N}$ by

$$A \in \widehat{N} \qquad \longleftrightarrow \qquad A = \sum_i B_i(a_i - \chi(a_i)) \, , \tag{2}$$

where $B_i \in \mathcal{V}_{\widehat{\mathcal{G}}}$, $a_i \in \widehat{\mathcal{N}}$. Now, we find all elements $C \in \mathcal{V}_{\widehat{\mathcal{G}}}$ for which the following condition is satisfied:

$$[X_n^\alpha, C] \in \widehat{N} \, . \tag{3}$$

These elements form the associative algebra V. And for the last "step"; we factorize V over left ideal $\widehat{N}$. This means that $C_1 \sim C_2$ if $C_1 - C_2 \in \widehat{N}$. It is easy to verify that

(A) $C_1 \cdot C \sim C_2 \cdot C$ if $C_1 \sim C_2$

(B) $W \stackrel{\text{def}}{=} V/\widehat{N}$ is the associative algebra. Let's call it W-algebra.

The statement: W-algebra contains the Virasoro algebra.

Indeed it is easy to verify that equation (3) is satisfied by the following elements of $\mathcal{V}_{\widehat{\mathcal{G}}}$

$$L_n = L_n^0 + K M_n \, , \tag{4}$$

where L_n^0 are the usual Sugawara expressions for the Virasoro generators, quadratic in currents, K is the central charge of the current algebra and

$$M_n \stackrel{\text{def}}{=} \sum_{\alpha \in R} (\alpha, \rho) M_n^\alpha \, . \tag{5}$$

Here R is the set of simple roots of $\mathcal{G}$, 2ρ is the sum of all positive roots and (α, β) is the scalar product of roots. It is easy to prove that in the same way as L_n^0, L_n generates Virasoro algebra but with another central charge.

In the case of $\widehat{\mathcal{G}} = \widehat{s\ell}(2)$, L_n generates all W-algebra. For $\widehat{\mathcal{G}} = \widehat{s\ell}(3)$ additional generators of the type

$$W_n = \sum_{\alpha,\beta,\gamma=1}^{3} \sum_{\substack{k,\ell,m \\ k+\ell+m=n}} J_k^{\alpha\beta} J_\ell^{\beta\gamma} J_m^{\gamma\alpha} + \quad \text{low order terms} \tag{6}$$

appear (here $(\alpha\beta)$ is the ordinary 3×3 matrix index).

In conclusion I want to point out the connection of this construction with the theory of integrable classical equations of KdV-type.

The KdV equation can be written in the form

$$\frac{d}{dt}\mathcal{L} = [\mathcal{L}, \mathcal{L}_+^{3/2}] \ , \tag{7}$$

where

$$\mathcal{L} \stackrel{\text{def}}{=} \left(\frac{d}{dx}\right)^2 + u(x)$$

and $(\mathcal{L}^{n/2})_+$ denotes the differential part of the pseudo-differential operator

$$\mathcal{L}^{n/2} = \sum_{k=-\infty}^{n} a_k(x)\left(\frac{d}{dx}\right)^k$$

(the coefficients a_k being defined through the recursive procedure from the relation $(\mathcal{L}^{n/2})^2 = \mathcal{L}^n$). The equation (7) is of Hamiltonian type. This means that on the space of functions $u(x)$ it is possible to introduce the structure of the phase space, i.e. the Poisson bracket and the Hamiltonian functional $H(u(x))$ such that equation (7) will take the form

$$\frac{\partial \mathcal{L}}{\partial t} = \{\mathcal{L}, H\} \ . \tag{8}$$

The remarkable property of (7) is the possibility to realize it by two ways. The first bracket has the form

$$\{u(x), u(y)\}_1 = \delta'(x-y) \tag{9}$$

and the second is

$$\{u(x), u(y)\}_2 = u(x)\delta'(x-y) + 2u'(x)\delta(x-y) + \delta'''(x-y) \ . \tag{10}$$

Certainly, the Hamiltonians for these cases are different. For the second bracket it equals $H = \int u^2(x)dx$. The integrals of motion $I_n^{(1,2)}$ of KdV equation produce the commuting flows on the phase space and are defined from the relation

$$\frac{\partial \mathcal{L}}{\partial t_n} = [\mathcal{L}, \mathcal{L}_+^{n/2}] = \{\mathcal{L}, I_n^{(1,2)}\}_{(1,2)} \ . \tag{11}$$

Long time ago it was noticed by Gelfand and Feigin that the second bracket is the quasiclassical limit of the Virasoro algebra. The Lax-Gelfand-Dikii procedure for KdV can be generalized if one considers an operator of the Nth order for $\mathcal{L}$:

$$\mathcal{L} \stackrel{\text{def}}{=} \sum_{k=0}^{N} u_k \left(\frac{d}{dx}\right)^k . \tag{12}$$

Then the generalized KdV equations have the form

$$\frac{\partial \mathcal{L}}{\partial t} = [\mathcal{L}, \mathcal{L}_+^{n/N}] . \tag{13}$$

These equations are also of Hamiltonian type and has two Hamiltonian structures.

It was recently noted by B. Feigin and T. Chovanova that for $N = 3$ the second Poisson bracket is the quasiclassical limit of the spin 3 W-algebra by Zamolodchikov.

In 1981 V. Drinfeld and V. Sokolov discovered the remarkable group-theoretical meaning of the second bracket. That is, they showed, that there is the KdV-type equation corresponding to each KM algebra (the ordinary KdV to $\widehat{s\ell}(2)$, etc.) and that the second Hamiltonian structure, connected with this equation, comes from the Kizillov-Beresin-Kostant Hamiltonian reduction with respect to the nilpotent subgroup (for $\widehat{s\ell}(n)$ this is the group of upper-triangular matrices). About the Hamiltonian reduction one can consult the text-book on mechanics by Arnold, appendix 3. I think that the construction described above is the quantum analog of the Drinfeld-Sokolov construction.

I am grateful to T. Onogi, A. Kato, K. Ito, S. Odake and T. Nakanishi for the help in the preparation of the manuscript and to V.G. Knizhnik for the translation of this talk. I also want to express my deep gratitude to Prof. K. Nishijima, Prof. N. Kawamoto and Tetsuya Onogi for their kind hospitality in RIFP during the '87 October meeting in Kyoto.

References

1. Zamolodchikov A.B. TMF, 1987, 65, p347 (in Russian)
Fateev V.A., Zamolodchikov A.B. Nucl. Phys., (1987) **B280**, 644 (FS18).
Fateev V.A., Lykyanov S.L., Landau preprint *The model of two dimensional conformal quantum field theory with Z_N-Symmetry.*

Part V

Related Topics

Some Remarks on Random Dynamics

H.B. Nielsen and N. Brene

The Niels Bohr Institute, Blegdamsvej 17, 2100 Copenhagen Ø, Denmark

1. Introduction

When a high energy physicist of to-day looks at his calendar he will see a long sequence of string conferences follow each other like pearls on a—string. And behind all these pearls is the belief that some string theory, superstring or heterotic string, may be the ultimate theory of everything, TOE. Credo?

One point should make us a bit sceptic: one of the virtues of the sacred string theories is that they contain the (sub)theory of gravity in a natural way, i.e. they perform essentially on a stage of the size of a Planck length, $1.616 \times 10^{-35}m$. Experiments of to-day test structures down to $\sim 10^{-18}m$, 17 orders of magnitude away from the Planck length. (A single exception is provided by the as yet negative outcome of the proton decay experiments which may have implications for features about 10^{-30}–$10^{-31}m$.) The job of testing string theories with the experimental high energy technique of to-day is therefore comparable to the job of testing the Standard Model of Strong, EM, and Weak Interactions with a technique able to distinguish lengths down to 0.1–$1m$ only, say a technique based on radio frequency waves and sound waves of low pitch or, largely equivalently, on the experience of the old greeks. Without doubt almost any feature of the standard model will be lost.

If we were able to tell a philosopher from the classic Greece that essentially all physical phenomena can be explained on the basis of the standard model with its built-in concepts of quantum theory, special relativity, gauge fields, strong, EM, and weak interactions, etc. he would certainly shake his head and think that we are completely crazy. For him it would be very unhealthy philosophy to base a theory on such concepts for which he has no evidence at all. He might occasionally have experienced materials, like amber or certain minerals, which attract each other weakly at small distances but should such phenomena really make him see the need for a complicated EM sector of the standard model instead of just saying that some materials attract each other? A mutual attraction may also explain why most materials hang together in lumps. He may see animal shells which all spiral the same way around or realise that he prefers to write with his right hand. Would that be enough for him to accept a theory with built-in parity violation in some interactions but not in others? Anyhow, the parity violation he might see probably has nothing to do with the violation in weak interactions. In summary, a classic greek philosopher—who has never heard about the physical experiments of the last few centuries or decades—would find the standard model with all its complicated mathematics completely superfluous if not genuine nonsense. Had he known Occams razor he would have applied it to the standard model.

Now, how should we look at string theories? Just counting orders of magnitude we should be as sceptic to strings as the old greeks to the standard model,

Springer Proceedings in Physics Vol. 31: **Quantum String Theory**
Editors: N. Kawamoto and T. Kugo

$$\frac{\text{string models}}{\text{LEP}} \approx \frac{\text{LEP}}{\text{classic greek physics}}.$$

(Here we use the symbol LEP to denote the $10^{-18}m$ physics of to-day, to be considered as Low Energy Phenomena on a Planck energy scale although being high energy phenomena on a human scale and requiring extremely fine accellerators.) One may, however, say that the theoretical physicist of to-day has the advantage of knowing all the steps that lead from $1m$ physics to the standard model and therefore is well equipped to make guesses in the direction of the Planck limit. We need a consistent theory of gravitation, and string theory is claimed to be a good candidate; string theory (in some of its versions) is also claimed to lead to the standard model in an extreme long wave length limit. The Planck limit, however, is as remote from LEP physics as LEP physics, i.e. the standard model, from 1m physics. Here we have tacitly assumed that the difficulty of extrapolating is proportional to the logarithm of the scale ratio. It might just as well depend on *log log(scale ratio)* or, worse, grow linearly with the scale ratio; in this case we may be completely lost.

To put string theory to a test one must make predictions or (most often) postdictions by going to some extreme limit and add extra auxiliary assumptions at intermediate steps. To obtain e.g. a group like $U_1 \times SU_2 \times SU_3$ (the standard model group) from the group $E_8 \times E_8$ of the heterotic string model [1] one must introduce assumptions about the way in which compactification takes place. This may give us the group E_6. To obtain the standard model group further assumptions on break-down mechanisms are needed.

It may be wise to examine which of the results obtained on the LEP level (if any at all?) do really reflect properties of the string model and which ones merely reflect the auxiliary assumptions introduced at the intermediate level.

One concievable outcome of such investigations may be the following: Almost all derivable features at the LEP level are consequences in a long wave length limit of auxiliary assumptions made on the intermediate levels, leaving almost no testable consequences of the original string theory (TOE). One may even imagine the following to be true: Almost all features at the LEP level are long wave length limiting consequences of some *not very specific* assumptions made on intermediate levels.

The existence of Yang-Mills theories at the LEP level is an example of a feature predicted by most string theories but also predicted by some σ-models [2].

We are now approaching the idea behind the *Random Dynamics* project [3,4]. In this project it is assumed (or hoped) that (almost) all features of LEP physics are consequences of limiting processes rather than consequences of any particular model (TOE) for physics at some very high level. Thus we could take any TOE from the set of all concievable TOE's, grind it through the machinery of the limiting processes and, if our philosophy is correct, at the end obtain the right LEP physics, i.e. the standard model. We might even let a computer generate a large set of randomly chosen TOE's to check whether all or almost all of them lead to the standard model.

That is not what we have done. Instead we assume that most of the concievable TOE's may be very complicated, so complicated that they appear very chaotic and can be treated as if they were random. Thus in principle we take a TOE constructed in a very general way and made as chaotic i.e. random as possible. One may e.g. take the space-time structure to be very complicated, not necessarily continuous, let the *action* of the theory depend on fields which take values on random manifolds, while single action contributions may have random coefficients and be invariant under randomly chosen groups.

In the ideal case the TOE should be made completely random, i.e. no basic *Laws of Physics* should be used as input. If never-the-less one could derive those laws of nature which are reflected by the standard model, merely as consequence of the guiding lines for the limiting process, our project would be completely successful. In practice we need as input some of the basic laws of physics and try to 'derive' some others, e.g. those which characterise the standard model.

Our results seem to indicate that many fetures of the standard model can be derived from a very random TOE, supplemented by a few of the basic natural laws. That may be taken as support for the point of view that we need not at all have a fundamental theory, at least not for the purpose of physics at the LEP scale.

We cannot exclude, however, that there really does exist a fundamental theory beyond the random level (Planck scale?), a theory which at the random level has developed some very complicated, chaos-like structure before it gives rise, in the extreme long wave length limit, to our relatively simple standard model physics. The few basic laws of physics serving as neccessary input or the guiding lines of the long wave length limiting process may then reflect properties of that fundamental pre-chaos theory and perhaps even require the existence of such theory.

2. Field Theory Glass and Gauge Glass as Models for Random Dynamics

The purpose of this section is to present a very general type of random models, the field theory glass, and a slightly less general type, the gauge glass.

The field glass model [5] is based on the following assumptions:

1. There is an underlying 4-dimensional Minkowski space-time, but
2. All physics is played on a stage consisting of a discrete subset $\mathcal{D}$ of points picked in a random but quenched way (i.e. taken randomly and held fixed in a given model or in a given calculation) from the set of all points in Minkowski space. The set $\mathcal{D}$ may be visualised as the instantaneous positions of the molecules of a perfect gas at equilibrium.
3. To each point i in $\mathcal{D}$ is associated some manifold $\mathcal{M}_i$ chosen in a quenched random way (i.e. the manifold changes from point to point).
4. To each point i in $\mathcal{D}$ is also associated a quantum field $\Phi(i)$ which takes its values in the corresponding manifold $\mathcal{M}_i$. Note that the field theory glass thus is a quantum theory. The field $\Phi(i)$ is a somewhat strange 'function' since it maps various points i on various manifolds $\mathcal{M}_i$. We may define a true function Φ by

$$\Phi : \mathcal{D} \rightarrow \bigcup_{i \in \mathcal{D}} \mathcal{M}_i; \quad \Phi(i) \in \mathcal{M}_i \; for \; all \; i \in \mathcal{D}. \tag{1}$$

5. There is an action S which has a very generalised form of locality; we assume

$$S = \sum_{i \in \mathcal{D}'} S_i(\Phi(j)); \quad \mathcal{D}' \in \mathcal{D}; \quad j \in Ball(\rho, i). \tag{2}$$

Each S_i depends only on the fields $\Phi(j)$ as defined on points j in a neighbourhood of radius ρ around the point i. We call this property *semilocality*. In the long wave length limit it leads to locality in the usual sense. The functional form of the contributions S_i is chosen in a quenched random way. We may say that each S_i is a quenched random function defined on the cross product

$$X_{j \in Ball(\rho,i)} \mathcal{M}_i. \tag{3}$$

It may be emphasized that the field theory glass thus introduced has a very general and random character: the space-time points in which the fields are defined, the manifolds on which they take values, and the functional form of the action, all are to be chosen in a quenched random manner. In contrast quantum mechanics, 4-dimensional space-time, and semilocality are input properties *assumed* to be true for the present model.

This model may in principle be translated to a computer program and one may again ask for properties in the long wave length limit, in particular properties which may appear like those of the standard model.

Our approach is more speculative and we are tentatively led to the concept of a gauge glass through the following steps

1. Within some region of $\mathcal{D}$ it may easily happen that all the contributions S_j to the action S are approximately invariant under some group $\mathcal{G}$ common for that particular region.
2. Via a mechanism described earlier [6] this approximate symmetry may give rise to an effectively exact gauge symmetry $\mathcal{G}$ for that particular region. The mechanism may be considered as the inverse of the Higgs mechanism. It may lead us to a field theory glass endowed with a kind of gauge symmetry, for which the group $\mathcal{G}$ is changing from region to region.

This is what we call a gauge glass [7], a model which has been the starting point for several previous investigations.

A gauge glass theory may also be seen as a generalisation of an ordinary lattice gauge theory by allowing a set of modifications:

1. The lattice is made irregular by letting the lattice sites take random positions in a quenched way. This corresponds to the property 2. of the field theory glass.
2. The link variables take values on a group space that varies from link to link in a quenched random way. This corresponds to the properties 3. and 4. of the field theory glass.
3. The plaquette contributions to the action are of quenched random functional form, corresponding to the property 5. of the field theory glass.
4. The gauge group may change from one region of the lattice to another and the way a given gauge transformation acts or is interpreted may change in a random quenched way from one site to another. As an example we may assume that the gauge group $\mathcal{G}$ has two isomorphic subgroups $\mathcal{G} = \ldots \mathcal{G}_1 \times \mathcal{G}_2$; $\mathcal{G}_1 \sim \mathcal{G}_2$; it is then possible that a given transformation may be ascribed to $\mathcal{G}_1$ at one site and to $\mathcal{G}_2$ at a neighbouring site. This property we call confusion.

3. Results

Most of the results presented here are based on the gauge glass model. They should be considered rather speculative. To check them would require major computer calculations as sketched above. Only one minor calculation has been attempted and the result was inconclusive [8]. We therefore prefer to formulate our results in a Darwinistic language saying that some property is healthy or fit for survival over the changes from the fundamental scale down (in energy) to the LEP scale, while other properties are unhealthy or unfit for survival. A gauge group is unfit if it is likely to break down by some collapse mechanism, such that some of its gauge bosons acquire masses on a scale far above the LEP scale. The group is also considered unfit if its Callan-Symanzik β-function is such as to give confinement at a high energy scale (large group, small number of fermions).

We have mainly concentrated about features which make a group vulnerable to collapse. There are several [9]:

1. The group has a trivial center (i.e. the center has only one element, the unit element). The reason why this property is unhealthy is the following: An extremum for the action (or energy) contribution from a single plaquette is most likely to occur at a special point of the group space, e.g. a center element. If there is only one center element, it is highly likely that many plaquettes have energy *maximum* at that element and consequently a noncentral energy minimum. That would classically lead to a spontaneous break down of the global part of the symmetry, in analogy to the Higgs model.

2. The space of conjugacy classes of the group has noncentral corners or other noncentral singular points. Again the reason for probable break down is that such points are likely positions for the extrema of the plaquette action.

3. The center of the group is disconnected. The danger of break down is due to a Bianchi identity to be fulfilled by the plaquette elements on the surface of a (hyper)cube (their product should essentially be the unit element). This identity is most probably not always fulfilled when the individuel plaquettes take their energy minima.The neccessary adjustment is likely to lead to an overall energy minimum for which many plaquette elements are not central, i.e. breaking of the global symmetry. A connected center is a healthy property, which may give the adjusted plaquette elements a good chance to remain inside the center.

At this stage one may note that the group behind the standard model is fit for surviving these dangers: Its center is certainly nontrivial (the center of U_1 is the U_1 itself and the centers of SU_2 and SU_3 have 2 and 3 elements, respectively); contrary to e.g. the SO_N groups, the SU_N covering groups have no noncentral corners or other singular points. There is *one* group (SMG) with the algebra of $U_1 \times SU_2 \times SU_3$ which has a connected center. (It is obtained from the covering group by division with a certain discrete subgroup of the center.) Moreover this group (SMG) has representations which fit exactly to the generations of quarks and leptons! We had one trouble with this group: how can it avoid having more than one U_1? The answer is that the occurrence of several factors U_1 is unhealthy due to confusion. Let us therefore continue the list of unhealthy properties for groups [7]:

4. The group has outer automorphisms. By the group of outer automorphisms we mean the (discrete) factor group obtained by dividing the group of all automorphisms of a given group by the subgroup of inner automorphisms. As mentioned in the preceeding section the way a given gauge transformation acts may be modified quenched randomly at each

particular site by an (outer) automorphism. (It is somehow like ordinary gauge, observers at different locations may use different coordinate systems for the internal variable.) In the simple case when the automorphism is a permutation of two isomorphic subgroups $\mathcal{G}_1$ and $\mathcal{G}_2$ we have explicitely shown that the group breakes down to a single $\mathcal{G}$ and one half of all the gauge bosons acquire masses. It was here assumed that the two sets of representations corresponding to the two subgroups have the same structure such that nothing can distinguish one group from the other. In this case we talk about collapse of symmetry due to confusion. Generalising this argument we get a special case of unhealthiness:

4a. The group has a multiple occurrence of some subgroup associated with the same multiple occurrence of a given set of its representations.

Outer automorphisms are not restricted to multiple occurrence of a subgroup. Some of the simple groups also have outer automorphisms, as can be inferred from their Dynkin diagram [10]. For the SU_N groups ($n \geq 3$) this outer automorphism is complex conjugation. The SMG is symmetric under complex conjugation; what saves it for survival is the existence of chiral fermion representations (quarks and leptons) which are not symmetric under complex conjugation. We can, however, still note a case of unhealthyness:

4b. The group is SU_N (or E_6) without chiral representations.

This completes our list of properties that might lead to collapse of the global part of a gauge group.

A recent application of the concept of confusion has led to a relation involving the possible number of quark-lepton generations [11]. The argumentation goes like the following:

At the gauge glass level there may be an n-fold occurrence of some group $\mathcal{G}$ (e.g. the SMG) accompagnied by the corresponding n-fold occurrence of some set of its representations. Confusion may lead to break down of this group $(\mathcal{G})^n$ into a surviving group $\mathcal{G}_{surv.}$ with only one set of massless gauge bosons. Some traces of the original group $\mathcal{G}$ are, however, left over: All the n sets of fermion representations transform in the same way under $\mathcal{G}_{surv.}$, just like generations behave under the standard model group at the LEP level. Moreover one can easily show that their common inverse squared coupling constant $\alpha^{-1}_{comm.} = (\frac{g^2_{comm.}}{4\pi})^{-1}$ is the sum of inverse squares of the coupling constants corresponding to the n original groups

$$\alpha^{-1}_{comm.} = \sum_{i=1}^{n} \alpha_i^{-1}. \tag{4}$$

By the mean field approximation one may calculate a critical value (upper value) $\alpha_{crit.}$ for on individual group. Insertion of this value in (4) gives a rough estimate for the number $N_{gen.}$ of generations

$$\alpha^{-1}_{comm.} \gtrsim N_{gen.} \alpha^{-1}_{crit.}. \tag{5}$$

This relation should apply at the gauge glass level ($\approx$ Planck level). It is supposed to be valid for the U_1, SU_2, and SU_3 couplings separately. Their values at the Planck level can be obtained from the values at the LEP level by extrapolation using the renormalisation group prescription. The final result is

$$N_{gen.} \lesssim 4 - 5. \tag{6}$$

A similar result has been derived using a Kaluza-Klein space-time model rather than the gauge glass model [12].

4. Conclusive Remarks

The results listed in section 3 are all based on the gauge glass model which was originally taken as a natural generalisation of lattice gauge models and in part inspired by the observation that the charge systematics of the fundamental fermions seems to point to a definite *group*, SMG, among the many groups sharing the algebra of the standard model. The gauge glass model may in turn be justified—through a kind of inverse Higgs mechanism—on the basis of a more fundamental field theory glass model. Our main result is that the standard model group SMG is among the fittest for surviving the dangers listed in section 3 down to the LEP level. That its algebra is among the fittest for survival may also follow from the following observations:

1. Solvable algebras [10] seem to play no rôle in Yang-Mills theories.
2. The 3 factors of the standard algebra, U_1, SU_2, and SU_3 have dimension 1, 3, and 8; these are the 3 lowest possible dimensions of non-decomposable algebras. The SO_4 algebra has only dimension 6 but SO_4 can be decomposed into $SU_2 \times SU_2$.
3. There may be good reasons for preferentially finding low-dimensional groups at the LEP level, e.g.
   - In a random world it would be more difficult for high-dimensional groups to arise originally.
   - High-dimensional groups may easier be subject to confinement and thereby disappear for us at a high level.
4. If we invoke a confusion mechanism it is understandable that U_1, SU_2, and SU_3 each appear only once and that SO_4 does not appear.

From these considerations one may conclude that most of our physical experience up to the LEP energy level tells us nothing or very little about physics at a fundamental level (Planck level?). Exceptions may be *Lorentz invariance* [13], *gravity* [14], and *quantum theory* [3]. Very ambitious side projects of random dynamics try to cover also these subjects, e.g. obtain features of quantum theory: hermiticity of the Hamiltonian and linearity of the Schrödinger equation, or derive quantum gravity on the basis of symmetry under diffeomorphisms.

To the extent that these latter projects are succesful the neccessity of inventing a definite TOE becomes extremely doubtful.

References:

1. Gross, D.J., Harvey, J.A., Martinec and Rohm, R., Phys. Rev. Lett. 54, 502 (1985); Nucl. Phys. B256, 253 (1985); B267 75 (1986)
2. D'Adda, A., Di Vecchia, P. and Lüscher, M., Nucl. Phys. B146, 63 (1978); B152, 125 (1979)
3. Nielsen, H.B., Bennett, D.L. and Brene, N., in Niels Bohr Centennial Conference, Recent Developments in Quantum Field Theory (North Holland, Amsterdam 1985); Weingarten, D., private communication.
4. Bennett, D.L., Brene, N. and Nielsen, H.B., Physica Scripta T15, 158 (1987)

5. Nielsen, H.B. and Brene, N., in Workshop on Skyrmions and Anomalies, Krakow 1987 (World Scientific Publishing Co., Singapore, 1987)

6. Förster, D., Nielsen, H.B. and Ninomiya, M., Phys. Lett. B94, 135 (1980); Schenker, S., Symmetry Breaking Operators in Gauge Theories (Research plan to DOE, USA, 1980)

7. Nielsen, H.B. and Brene, N., in Proc. XVIII Intern. Symp. Ahrendshoop, Institut für Hochenergie physik (Akademie der Wissenschaften der DDR, Berlin-Zeuthen, 1985).

8. Hansen, K., Dissertation (Niels Bohr Institute, 1986).

9. Nielsen, H.B. and Brene, N., Nucl. Phys. B224, 396 (1983).

10. Gilmore, R., Lie Groups, Lie Algebras and some of their Applications (John Wiley & Sons, New York, 1974)

11. Bennett, D.B., Nielsen, H.B. and Picek, I. in Proc. XX Intern. Symp. Ahrenshoop, Institut für Hochenergie physik (Akademie der Wissenschaften der DDR, Berlin-Zeuthen, 1985).

12. Bennett, D.B., Nielsen, H.B., to appear in Proc. Beijing Workshop on String Theories, China, 1987

13. Chadha, S. and Nielsen, H.B., Nucl. Phys. B217, 125 (1983) and ref. therein

14. Lehto, M., Nielsen, H.B. and Ninomiya, M., Nucl. Phys. B272, 213 (1986); B272, 228 (1986)

Index of Contributors